P9-CFB-500

PRINTER'S ERROR

PRINTER

ERR

PRINTER'S ERROR

IRREVERENT STORIES FROM BOOK HISTORY

J. P. ROMNEY AND
REBECCA ROMNEY

An Imprint of HarperCollins*Publishers*

HarperCollins books may be purchased for educational, business, or sales promotional use. For information, please email the Special Markets Department at SPsales@harpercollins.com.

FIRST EDITION

Designed by Bonni Leon-Berman

Library of Congress Cataloging-in-Publication Data has been applied for.
ISBN 978-0-06-241231-7

17 18 19 20 21 LSC 10 9 8 7 6 5 4 3 2 1

Rebecca dedicates this book to Richard D. Altick and Elizabeth L. Eisenstein, neither of whom would have entirely approved.

J.P. dedicates this book to the Ghost of a Flea, for being the coolest thing he's ever seen. He and Rebecca also dedicate it to Ellie and Kit, who are pretty cool, too.

Posterity it seems has to soften and make respectable, smooth and polish, unable to see that the rough, the raw, the discordant, may be the source and nurse of creativity.

—*Doris Lessing*

Transgression is one of the historian's most useful data, for the pointed figure of the scandalized is often our index of the norm.

—*Joseph Loewenstein*

CONTENTS

INTRODUCTION
WHAT DO YOU RECKON THIS IS?

THIS IS AN IRREVERENT HISTORY. Sure, to most of us, a history about the printed word sounds dignified. After all, the printing press has recorded and spread some of the greatest achievements of humankind. But remember, humankind is also full of idiots. Ridiculous decisions, weird coincidences, and absurd stories are no strangers to our species. The history of the printed word reveals our capacity for brilliance, but it also reveals our capacity for blunder. The printing press is a stage upon which the entire drama of human thought and morality is acted out. At times that's the delight of Beatrix Potter's *The Tale of Peter Rabbit*; at others, it's the adamantine weight of James Joyce's *Finnegans Wake*. Sometimes the ideals are as enlightened as John Milton's *Areopagitica*, and at other times they're whatever the hell *Fifty Shades of Grey* is.

Human civilization is all the richer for the bizarre history of printed books. Our mistakes, our provocations, and our mysteries are all valuable in their own ways. Our predecessors who lived hundreds of years ago are not so different from us, and we can't help but laugh in recognition of our shared folly. Take the 1631 printing of the Bible in English, the infamous "Sinners' Bible." The printer, Robert Barker, owned the exclusive privilege to print Bibles in London. His greatest triumph was the first edition of the King James Bible (1611). If you're able to find one today, you can expect to pay one hundred to two hundred thousand dollars (or more) for it. Near the end of Barker's career two decades later, he produced one thousand copies of the Holy Book without a very close proofread. While most typos are innocuous, others lead to trouble with the law. In setting the type for Barker's sinful Bible, a compositor somehow forgot a key word in the seventh of the Ten Commandments. "Thou shalt not commit adultery" became "Thou shalt commit adultery." When this "mistake" was caught, Barker and a colleague were summoned to a legal hearing at the order of King Charles I. They escaped with only a hefty fine, but copies of the Bible were seized wherever they were found, and destroyed. Partially because of the scarcity, today a copy of this "Wicked Bible" can fetch well into the tens of thousands of dollars at auction. Modern collectors appear as amused by the error as the the eighteenth-century owners who went out of their way to save their copies for posterity.

People like to push, to test boundaries. Scandal and amusement are two of the most glorious strands woven into the tapestry of the printed word. Take one of the masterpieces of American children's literature, *The Adventures of Huckleberry Finn*, by Mark Twain. Published in 1885, Twain's masterpiece has everything a reader could ever want: adventure, humor, controversy, tenderness, the hypocrisy of competing social class structures—and a penis.

Most people agree that exposed adult genitalia are not something you should normally encounter in children's literature. Yet according to the first printed copies of *Huckleberry Finn*, Uncle Silas did not agree with most people. Mark Twain didn't intend his farmer/preacher character to be a sex offender, but in the weird world of print, mistakes (or, in this case, pranks) happen. During the printing, the innocent illustration of Uncle Silas was altered without Twain's knowledge to appear as if Silas were proudly displaying his erect phallus—a phallus that, in order to pass unnoticed at a casual glance, had to be the size of a mealworm.

"WHO DO YOU RECKON IT IS?"

"WHO DO YOU RECKON IT IS?"

"WHO DO YOU RECKON IT IS?"

THE INFAMOUS UNCLE SILAS ILLUSTRATION IN ITS THREE STAGES: THE "CURVED" FLY (BEFORE), THE DEFACEMENT, AND THE "STRAIGHT" FLY (AFTER). *Image used with the kind permission of Kevin Mac Donnell Rare Books.*

Despite a five-hundred-dollar reward (more than twelve thousand dollars in today's currency), we still don't know exactly which of the fifty pressmen working on *Huckleberry Finn* vandalized the printing plate. In the original illustration, which depicts Huck Finn presenting himself in Uncle Silas's home as his distant nephew, Silas stands with his pelvis jutting out in an exclamation of dramatic confusion. Aunt Sally hovers nearby with a half-smile, as if she suspects what's really going on. And young Huck has his hand on his hip, thinking of how to respond to the old man's line of questioning.

With a few scratches into a printing plate, however, the scene took on a markedly different feel. In the altered sketch, Aunt Sally has a disturbing grin on her face, Huck appears to be silently processing the obscenity, and Uncle Silas looks an awful lot like he is gesturing to his tiny exposed member over a caption that reads: "Who do you reckon it is?"

One might suppose that the "greatest American humorist" would have seen the lighter side of this little printing debacle. Mark Twain did not. The famed author immediately had the illustration stripped from all copies, and book agents selling subscriptions door-to-door were ordered to tear out the page from their display copy. Although no copies containing the hilariously offensive Uncle Silas illustration made it into the hands of the public at large, some agents inevitably bucked Twain's orders, and the drawing ended up displayed in the offices of a few publishers.

Fixing the penis debacle cost Twain a tremendous amount of money and delayed the book's release beyond the 1884 Christmas season. Today we read the phrase "published in 1885" and don't realize the small shit storm that changed the book's publication date from 1884. Modern rare book dealers always note the state of Uncle Silas's fly in their descriptions of the first-edition copies of *Huckleberry Finn*, which sell for thousands of dollars.

We love exploring the darkly comic lining to history. Admittedly, this can sometimes be a challenge. For example, here is an iconic line from Dostoyevsky's masterpiece *Notes from Underground*: "I'm certain that man will never renounce real suffering . . . why, this is the sole cause of consciousness." Russia is not a funny place. Just ask the Soviet-born comedian, Yakov Smirnoff: "Many people are surprised to hear that we have comedians in Russia, but they are there. They are dead, but they are there." If Smirnoff can find humor in the USSR, we figure finding humor in the history of the printed word can't be *that* hard, right? Whether it's a fifteenth-century German inventor short on money or a London engraver talking to ghosts three centuries later, people are people, and part of being people is doing bat-shit crazy things.

Take that inventor from Mainz. Nearly everyone recognizes the name Gutenberg, but what do we really know about him? Shockingly little. Gutenberg may have invented the printing press in the West, but he didn't print anything about himself. In fact, for hundreds of years, the European inventor of one of mankind's greatest innovations was unknown to the world. In contrast to these painful gaps in the historical record, we have a wealth of material about one Johannes Trithemius, a Benedictine monk who campaigned *against* the new invention of printing three decades after Gutenberg. Ironically, we know so much about this obstinate monk because others *printed* his arguments against printing. Who gets to determine how we are remembered after we are dust?

Contradictions like this come alive in the history of books. Gerardus Mercator, a sixteenth-century Flemish mapmaker who coined the term *atlas*, is the man responsible for making it possible to use a flat map to circumnavigate a round globe. Yet he's also responsible for generations of Europeans believing the Arctic was dominated by a ring of four super islands inhabited by "Little People." Mercator added that tidbit on good authority, or so he believed.

When faced with a flood of information, misinformation, and conflicting sources, how do you know what to trust? The books in front of us seem so pure and pristine, but like the surface of a quiet lake, underneath those printed words are murky depths. As with Gutenberg, most people know Shakespeare, but most people *don't* know that there are "bad" versions of his plays out there. Hamlet may be one of the most eloquent characters of Western literature, but the earliest printed version of *Hamlet* makes the Danish prince sound like a drunken pirate at an English Christmas party: "To be, or not to be, Aye there's the point." When you dive deeper, you start to wonder if we really know Shakespeare after all.

The world of print is full of ironies, too. Take the greatest translator of the English Bible, William Tyndale, whose work made the Holy Book accessible to countless worshipers for the first time—and who was burned at the stake for heresy. Or think of the husband of the protofeminist Mary Wollstonecraft, who published a devoted memoir thinking it would ensure his wife's reputation—only to see it push her into public shame and obscurity for the next hundred years. Who is allowed to speak as an authority when the printing press democratizes power and encourages the widespread discussion of ideas?

The passions and foibles of human nature are exemplified in what we print and why. Benjamin Franklin formed an empire of print that would go on to help unite an emerging nation. (This empire also made him an ungodly amount of money.) When the British Parliament levied a now-infamous tax on paper, they didn't realize that printers, with their ready-at-hand soapboxes, are the last people you want to piss off. One adage was as true then as it is today: follow the money.

Through it all, we've made some exquisite books. Yet even in the best of them, our flaws cannot be separated from our triumphs. William Blake was a visionary artist and engraver as well as a poet, responsible for a clever new illustration technique—an innovation he claimed to have learned from communicating with the dead. The bookbinder Thomas Cobden-Sanderson helped create the Doves typeface, considered (by us) to be the most beautiful font in the world, then plotted to destroy it by sinking his masterpiece into the Thames.

Make no mistake, books are a reflection of us. Just watch the wild battle unfold between Charles Dickens and the entire United States of America as we declared in no uncertain terms our right to plunder other people's work. In the nineteenth century, to err may have been human, but to steal was as American as apple pie, or baseball, or the Statue of Liberty. (Actually, none of these things were considered "American," at the time, so just the stealing, then.)

Books printed one hundred years ago or more still have much to teach us about our own world. By the 1920s mass consumerism had cemented itself in the United States with the help of ad men who took the momentum of selling tobacco to women as a form of weight loss and turned it toward convincing Americans to buy more books.

How do we reconcile our respect for books, those powerful symbols of our intellectual heritage, with our less virtuous predilections?

The flawed history of humankind can be found in books, and the questions books provoke still drive us today. The printed word is glorious, but it's also nuts—because we are gloriously nuts.

If you're reading this, you love that fact as much as we do. The book historian Richard Altick once wrote, "There has always been a popular belief that more than casual attention to books is either a symptom or a cause of madness." That means what the Cheshire Cat said to Alice could well be said of the reader who picks up this book:

> "But I don't want to go among mad people," Alice remarked.
> "Oh, you can't help that . . . we're all mad here. I'm mad. You're mad."
> "How do you know I'm mad?" said Alice.
> "You must be," said the Cat, "or you wouldn't have come here."

Welcome to the madness of printed books. Let us show you around.

PRINTER'S ERROR

ONE CHAPTER ONE 1

HOW TO FORGE A RARE BOOK

IN JUNE 2005 AN ITALIAN book dealer named Marino Massimo De Caro presented for sale a proof copy of Galileo Galilei's first telescopic observations of the night sky. This publication included Galileo's latest discoveries about the moon, never-before-seen stars, and the previously unknown moons of Jupiter. Galileo's signature was included, and the original watercolors. It was a copy to set your heart racing.

A proof copy of *Sidereus Nuncius* (*The Starry Messenger*), originally published in March 1610 and regarded as one of the most important technical treatises produced by the human race, promised to be a major event in the history of science. De Caro easily found an interested buyer, the rare book dealer Richard Lan, who partially specializes in landmark works from the history of science. Yet for a book of this magnitude, Lan knew that authenticating the volume with due diligence extended beyond even his decades of experience. He brought in some of the world's most respected scholars in early printed books, all eager to determine whether this copy was as game changing as they hoped.

Over the summer of 2005 the professionals conferred, the book was inspected, and a seal of approval granted. Then came the most thrilling moment of all: the deal. De Caro agreed to let his potentially ten-million-dollar copy of *The Starry Messenger* go for the bargain basement price of half a million dollars. It was "the acquisition of a lifetime," Lan said.

There was one tiny catch: the treatise was fake. Beginning to end, paper to ink, De Caro had fabricated the whole thing. Over the centuries, rare book specialists have contended with things like facsimile pages, restored bindings, and pages supplied from other copies, but an *entire sixty-page treatise*? Forging a book of this magnitude is so unheard of that, for seven years, experts and appraisers thought the Holy Grail of scientific history had simply dropped from the sky.

Nothing illustrates the messiness of print quite like forgery. Forgers are one part intellectual, one part psychopath, equal doses narcissist and megalomaniac, with a lemon twist of hydrochloric acid thrown into the mix (more on that later). Their motivations aren't always the same, but every decent forger has one thing in common: a wide knowledge of history. Without that, they don't last very long.

Forging an entire book is an astronomically difficult endeavor. Simply put, there's just too much specialized knowledge for one person (or a small army of persons) to keep track of. The forger would have to master dozens of different disciplines, from the paper, to the ink, to the binding, type, press, and illustrations (most of them saturated with more than five hundred years of quirks and technicalities), to successfully fabricate a book. One slipup, anywhere along the way, and the deception is revealed.

Run-of-the-mill forgers are more likely to set their sights on easier pursuits, such as single documents, sports signatures, or robbing the Federal Reserve. But for every crime with a horrible chance for success, there's always someone out there willing to give it a try.

Sometimes these attempts end up as hilarious viral videos. Other times someone actually gets away with selling a fake copy of Galileo's seventeenth-century *Starry Messenger*—for a time, anyway.

What sets De Caro's *Sidereus Nuncius* apart from most other forged books (including similar attempts, such as De Caro's forgeries of Galileo's treatises on the geometrical compass) is the sheer magnitude of the ruse. Traveling between Italy and Argentina over a period of two years, De Caro and a handful of alleged conspirators meticulously handcrafted their very own seventeenth-century Venetian imprint. When the fake was finally uncovered in 2012, the word *masterpiece* was thrown around by some of the experts who were first fooled. Other scholars, for their part, would be embarrassed that it took so long to pull back the curtain on De Caro's wizard.

Fabricating a rare book is like forging a world-famous painting . . . if you also have to re-create the museum it's housed in, right down to historically accurate light fixtures and doorknobs. Traditionally, most rare books have been much more expensive to forge than they are actually worth. This is partly why De Caro wasn't immediately caught. Scholars weren't looking for a forgery. Who in their right mind would go to such great lengths to fake a whole book? A criminal, it turns out, who thought he was smarter than everyone else in the rare book world, and was willing to put this belief to the test.

BORN IN Italy in 1973, Marino Massimo De Caro is, by many accounts, a very likeable guy. He is tall and heavy-set, with blue eyes and a face that could pass for that of either a kindly book dealer or, with a fedora and furrowed brow, an enforcer for the Corleone family. De Caro entered politics at the ripe age of twenty-two, and even those who later would be swindled by him have described him as a socially gifted young man.

De Caro's involvement with rare books took off at age twenty-five, when he dropped out of law school, took a menial job working in a pension office in Verona, and began frequenting the shops of local book dealers. The next year he was traveling to international book fairs as a collector. Four years later, he was brokering book deals worth over a million dollars. Two years after that, he offered his fake copy of *Sidereus Nuncius* for sale.

De Caro's choice of Galileo was anything but serendipitous. He claims that, ever since he was a boy, his admiration for the seventeenth-century Father of Modern Science bordered on obsession. While most American teenagers in the 1980s were reading a mixed bag of *The Scarlet Letter*, *X-Men*, and *Playboy*, De Caro boasts he was sitting in a library picking his way through 4,200 pages of Galileo's known letters. In 2007 he would publish a two-volume biography of Galileo, a labor of love that William Shea, a Galileo expert, would later compare to "an extended undergraduate paper with no quotations—the kind of thing an American student would pull off the Internet."

An astronomer who lived three hundred-plus years earlier became De Caro's moral compass. De Caro saw Galileo as a nonconformist and a rebel—the original rebel, really. James Dean, with his tousled bouffant and head-on collisions, had nothing on Galileo, with his receding hairline and peaceful death in bed from heart palpitations. In De Caro's estimation, Galileo was a giant among men—which, coincidentally, was also Galileo's estimation of himself.

Don't get us wrong: Galileo is worthy of admiration. He was a visionary who had to fight to convince people of the realities he observed through his telescope. His support of Copernicus's theory that the sun does not revolve around the earth is one of his more famous fights. It is worth noting, too, that during his struggles against the Catholic Church, Galileo called his priestly critics "mental pygmies," "dumb mooncalves," and persons "hardly deserving to be called human beings." This could only have endeared him more to the teenage De Caro. As a brilliant man who was "vehemently suspected" of heresy, Galileo became De Caro's patron saint of Science and Rebellion and Sticking It To The Man. "I wanted to use the philosophy of Galileo *against* [the scholars]," De Caro claimed to *New Yorker* reporter Nicholas Schmidle after De Caro's crimes were uncovered.

Defending forgery, embezzlement, and petty larceny by pointing to Galileo's lifelong struggle to drag the Catholic Church kicking and screaming into the Scientific Revolution is a weak justification for his crimes. Nevertheless, De Caro tried to position himself as a modern-day Galileo, fighting his own "mental pygmies"—who, in his mind, constituted all the scholars alive in the world today. Instead of blasting them with a groundbreaking scientific treatise, however, he would boldly fake *someone else's* groundbreaking scientific treatise and sell it at a fraction of its current market value. The fools!

Let's not kid ourselves. It was more likely about the money.

As the director of the Biblioteca Girolamini in Naples, and using connections he made with libraries all around Italy, De Caro began systematically stealing authentic imprints by Galileo and other writers. His thefts initially went unnoticed because he replaced these authentic editions with his forgeries. This strategy allowed him to sell real Galileos without risking his fakes on the open market.

And that market was ravenous. In recent years, collectors have craved the "high spots," that is, the best-known and most beloved books within any given subject. For literature, this means books such as Ernest Hemingway's *A Farewell to Arms* or Jane Austen's *Pride and Prejudice*. For exploration, it means the accounts of Lewis and Clark's expedition or Captain Cook's voyages. In the past twenty years, major works in the history of science have jumped in popularity in the rare book world. To misquote the esteemed fashion mogul Jacobim Mugatu from *Zoolander*, "Science, so hot right now. Science." Copies of the first edition in English of Newton's *Principia Mathematica*, the work that first outlined the laws of motion and gravity, were being sold by dealers for around twenty thousand dollars in the late 1990s. Yet comparable copies were being sold by those same dealers in the early 2010s for around seventy-five thousand. Not every book has seen such an incredible jump; in fact, it's rather unusual. Yet when you combine two major collecting trends (in this case science and high spots), the market loses its goddamn mind. Thus, to a rare book dealer, De Caro's choice of Galileo as the subject of his forgeries reads more "criminal trying to get the most bang for his buck" and less "Galileo is my hero, OMG, I want to be just like him!"

When forging a book such as Galileo's, one of the most obvious places to start is the paper. De Caro knew this, and reportedly studied papermaking by hand in Italy. He took this knowledge to Argentina, where he enlisted the help of a book dealer—allegedly; the case still hasn't been brought to trial—and a local artist. Together they manufactured authentic-looking sheets of "rag paper" to use in the Galileo forgery.

Four hundred years ago, paper was made from rags (or, technically, the cloth fibers that make up rags). Rag paper sounds like something cheap, but in fact, it's quite the opposite. In comparison to what we use today, rag paper was downright luscious, often thick and soft to the touch. This is one of the main reasons so many early printed books are missing blank leaves: owners would cut them out to use them for other purposes. *A perfectly good piece of expensive paper, just sitting there not being used? What a waste!* In the world of rare books, many collectors will still consider buying a fifteenth-century printed book that is missing blank leaves. That's not the case for incomplete books from the nineteenth-century; collectors flee from them like they've just seen It peering out of a storm drain.

Each time period has its own rules for collecting. What's okay in one era isn't necessarily okay in another.

Back in Galileo's time paper was so expensive that, when printing a book, the cost of the paper could easily equal the cost of everything else *combined*, including the labor of the compositors and pressmen. The cost of paper varied depending on the quality of the cloth rags broken down and reformed into paper. The nicer the rags, the nicer (and more expensive) the paper.

Scarcity of supply was a constant problem for paper manufacturers. The sixteenth-century astronomer Tycho Brahe, who founded a paper mill for printing his own books, requested donations so often that his pleas were nicknamed "rag sermons." As the need for paper grew, men, women, and children of the lower classes were often employed as rag pickers, making their living sifting through refuse piles to find bits of cloth to sell to the paper mills. If you happened to be employed as a young rag picker (and we use the term *employed* here loosely), you would be keeping your seven-year-old eyes open for linen cloth, the diamonds of the garbage heap.

Clothing in the seventeenth century was made primarily of wool (from sheep) or linen (from the fibers of the flax plant). Wool fibers were not ideal for paper production because they gummed up and matted together when wet. Linen worked quite well, though. This meant that paper mills could transform old stockings and used underwear into books, newspapers, and treatises that revolutionized the very foundation of scientific understanding. This is one of the reasons that paper from the first few hundred years of print can feel so delectably supple: its texture resembles that of fabric (read: your favorite pair of underwear) more than the paper we use today.

Just like most forgers, De Caro would have known that 1844 marks an important date in the development of paper production. That was the year when wood pulp was successfully added to the industrial production of paper. Wood was much easier to obtain, so its inclusion made papermaking enormously cheaper. Of course, the paper feels different as a result. Add sawdust to a loaf of bread, for example, and it won't "taste better" or "be palatable in any way," but it will be more economical to produce. Replace linen rags with wood pulp, mix in a healthy dose of the Industrial Revolution, and voilà!, you get a cheaper, mass-produced, gross-tasting paper. And yes, we learned that last tidbit from experience: not by eating paper ourselves, but by observing which books have been eaten by something else.

Allow us to explain.

Wormholing is the term that bookdealers use when describing paper damage caused by book-boring insects. These "bookworms" like to munch tunnels through nice, clean cloth, but for some reason have an aversion to chemically bleached industrial wood pulp. Unsurprisingly, wormholing can be found commonly enough in books, especially those from the fifteenth and sixteenth centuries, but by the nineteenth century it pretty well disappeared. While evidence of wormholing isn't exactly *preferred* by collectors, it rarely hurts the value of the book in a significant way.

Not only did bookworms take a blow from the introduction of wood pulp to paper, but so, too, did those little ragpickers from the refuse dumps. "Picking" took a downturn, and countless trash-sorting children sadly found themselves unemployed. (Cheer up, kids, you're just in time for the Industrial Revolution, when there will be plenty of bone-grinding machines that need tiny fingers to clean them.)

The introduction of wood pulp inevitably changed the tactile sensation of paper. If you were holding an 1885 first edition of *Huckleberry Finn*, for example, the pages would feel noticeably stiffer than the pages of an 1818 first edition of *Frankenstein*. The paper used in the latter would be rag-based, and therefore softer to the touch. If you're a collector, such details are all part of the joy of discovery.

For a forger, however, they're a nearly insurmountable challenge. One of De Caro's other Galileo forgeries was uncovered in part by the out-of-place stiffness of his paper.

It would also be hard to miss the difference in smell over time. Think about the oldest book you own. For most people who are not rare book dealers, this would possibly be a beloved childhood copy of R. L. Stine's *Goosebumps* that you found in the back of your closet. You slowly crack open those pages, and you can just smell how great the paper is. Such a wonderful moment. This is how a book is supposed to smell, you think. All nostalgic and well worn and luxurious.

Well, you're wrong. You're not smelling luxury when you crack open those aged pages; you're smelling the inherent cheapness of paper. Wood pulp contains an aromatic organic polymer called lignin, which gives paper made after 1844 a vanilla-like smell over time. While this aroma is one of the single most pleasant experiences to a modern-day bibliophile, for most of the history of print that smell did not exist in books because they weren't composed of materials with a high percentage of lignin. You didn't want pretty smells in your books. You didn't want *any* smells in your books. Besides the scent produced by lignin, odors are most often indicative of dirt, mold, or other unpleasant things an elementary school student might rub into her Scholastic copy of *Goosebumps*.

At least by the time he forged *Sidereus Nuncius*, De Caro was aware of the major pitfalls of fabricating seventeenth-century paper. He would have known that wood pulp was a dead giveaway, so, instead, he selected handcrafted rag paper. But the details always get complicated when it comes to the history of print. The definition of the word *rag* in papermaking has changed a bit over the past four hundred years, and this little fact became one of the most damning pieces of evidence against De Caro. The rag of Galileo's day was made mostly from bast fibers, the basis for linen fabrics. The rag of today is most commonly made of cotton linters. Having cotton in your Galileo forgery doesn't necessarily reveal the fraud, as both cotton and hemp were used to a degree in paper production at the time. But cotton *linters* are a smoking gun. As they might say on *CSI: Miami*, "That's why De Caro wasn't able to [puts sunglasses on] pull the cotton wool over everyone's eyes for long." (That will be the last pun in our book. That is cotton-based.)

Cotton linters are the thin, velvety fibers that remain stuck to the cotton seed after the ginning process. Separating cotton from its seeds manually has always been difficult and time-consuming, so the 1793 invention of the modern mechanical cotton gin was a major breakthrough for the papermaking industry. The machine used a roller combined with a sort of grid through which metal spikes pulled the cotton fibers free from the seeds. Unfortunately for De Caro, the cotton gin was invented one hundred eighty years after Galileo's *Sidereus Nuncius* was published. Worse still, the machines used to separate linters after ginning weren't even available until the nineteenth century. These silky cotton scraps are a primary fiber source for making rag paper today, but they are anachronistic red flags in any book supposedly printed in 1610.

Sometimes we see what we want to see, however. With a little pigment coloring, a spell in a kitchen oven at 480 degrees Fahrenheit, and a dash of dirt rubbed into the corners and gutters of the pages, De Caro "aged" his paper convincingly enough that experts initially overlooked the presence of the cotton linters. Later inspection would reveal the rather elementary faux aging techniques that De Caro used. Even invisible fingerprints left by the forger were discovered, glowing brightly under UVA light.

For anyone to accept his forgery, De Caro knew that he would have to create a convincing and functional printing ink. Johannes Gutenberg came to a similar conclusion five and a half centuries earlier, though with presumably less criminal intent or Italian swagger.

Inks had been in use for thousands of years before Gutenberg's press. Ingenious methods had been developed to add lasting pigment to inks, which included the color sepia extracted from cuttlefish, purple extracted from mollusks, gold from fish bile, red from cinnabar ore, blue from indigo plants, and yellow from saffron flowers. The black that went into writing inks came mostly from soot or iron salts.

In the Europe of Gutenberg's day, iron gall inks (used primarily as writing ink) could be made only with the assistance of the female gall wasp. In order to reproduce, the gall wasp must lay her eggs into a tree, most often oak. A fairly mysterious process takes place wherein the site of the injection swells, filling with nutrients for the larva inside, which will break free from the hardened protuberance (the *gall* or *nut*) when it's fully developed. (The scientific term for this is not "busting a nut," but c'mon, it should be.)

Ink makers collected these unbusted oak galls, ground them up, and soaked them in water to extract the tannic acid, which was mixed with iron for coloring and bound with a vegetable gum. As widespread as iron gall was at the time of Gutenberg, however, writing inks were not ideal for use on a printing press. For one thing, the tannic acid was unsuitably strong. (Fans of *Alien* know that acid and metal are not friends under any circumstances.) Over time, the iron gall would eat away not only the metal type of a press, but also the fibers of the paper.

Binding properties were another problem Gutenberg faced with water-based inks such as iron gall. Printing ink has to be sufficiently fatty, or it won't stick to the metal type. Bucking the popular inks of his day, Gutenberg created his own formula, a recipe whose precise makeup, incidentally, remains a mystery today. Recent tests suggest that his secret formula may have been most closely related to oil paints used by local artists.

Over the years, testing of antique inks has uncovered some fairly strange additives—for example, human urine. Why would you ever find urine in ink? Because the large leather balls used to transfer ink to the metal type of the press (looking not unlike comically oversize candied apples) had a tendency to become stiff and cracked over time. And what's the best way to maintain the suppleness and youthful appearance of leather? Soaking it in pee. Apparently.

As he was creating the first printing ink, Gutenberg realized that vegetable oil would provide the fatty substance necessary to allow printing ink to stick to metal type. Along with the press itself and movable type, this ink was Gutenberg's most important innovation, one that allowed for the success of his printing press in Europe. Linseed oil (expressed from flaxseed) was the primary vehicle for antique printing inks. Colored with carbon soot, this rich black ink served the world of printing since the 1400s. Not until the twentieth century did petroleum-based inks gain traction.

Because the basic composition of printing ink changed so little over the first five hundred years of movable type, De Caro was able to hide his Galileo forgery within a *generous* margin of ink error (think of kids playing hide-and-seek with the boundaries set as all of North America). It was virtually impossible for researchers to determine if De Caro's book had been printed with fraudulent ink. Tests do exist that can give a clearer picture of an ink's age (e.g., C-14 analysis), but those tests require invasive procedures. When you've potentially got the world's most breathtaking new discovery in the history of science in your hands, you tend to avoid slicing off pieces of it. Thus the printing ink was taken at face value, without further testing.

But De Caro still had to contend with an additional kind of ink for those fake Galileo watercolors—which would be scrutinized by the world's most respected art historians. De Caro claimed that, with the help of an alleged accomplice in Argentina who had a background in pharmacology, he began testing the acidity levels of different inks available on the open market. For the watercolors, De Caro chose an India ink dating back to the 1800s that, when compared to modern inks, tested acceptably high on acid. Following a line of thinking that could be described in similar terms, De Caro resolved to find a way to age that ink artificially by two hundred years, transforming his unremarkable India ink into seemingly authentic seventeenth-century Galileo watercolors. And like many a harebrained alchemical scheme, his work started in the kitchen. With hydrochloric acid.

De Caro says he placed the forged pages, with their dark, fresh ink, on the top rack of a totally ordinary oven, then inserted a glass baking dish on the bottom that contained hydrochloric acid. Cranking the oven up to 480 degrees, he cooked the pages. The rising fumes from the acid oxidized the fresh ink, transforming it into a toasty sepia brown. In a kind of reverse Fountain of Youth, twenty minutes of roasting acid is equivalent to about four hundred years of natural aging. In the end, De Caro achieved the perfect marshmallow finish to his paper, and no one could easily dispute that the ink on his pages wasn't placed there in Italy in the year 1610.

After this hydrochloric sleight of hand, De Caro might have gotten away with his forgery if it hadn't been for those meddling kids! Nope, wrong forgery case. De Caro might have gotten away with it if he hadn't tried to fabricate an entire book, complete with Galileo's signature, original illustrations, binding, and replica stamp from the personal library of Federico Cesi (Galileo's patron and founder of one of the most important scientific fraternities in seventeenth-century Rome). With so many variables, De Caro was bound to make a mistake (well, many mistakes actually) along the way.

That's not to say he didn't take clever shortcuts. Rather than fabricate the book binding from scratch, he found one that already existed. Also, he nestled his forgery among authentic Galileo works, all of which were subsequently bound together as a Sammelband in an actual seventeenth-century vellum binding. This is the literary equivalent of a turducken: a forgery inside an anthology inside a genuine leather cover.

Minute differences in sewing techniques and threads would later form some of the most unequivocal evidence proving the fraud. Initial investigations, however, took for granted that the binding was authentic to the time period. Just as De Caro had hoped. After all, why go to the effort of fabricating "antique" leather and cardboard when you can let some poor schmuck from the seventeenth century do it for you?

Back before humanity discovered cardboard (an important date to mark on your calendar, along with other man-made materials such as plastic, neoprene, and whatever the hell is in chicken nuggets), the boards forming the front and back covers of a book were made from heavy, relatively expensive slats of wood. It's amazing to handle a book with wooden boards: the weight in your hands works as a sort of sensory subtext, making the book seem that much more impressive. But after the explosion of print in Europe, these wooden covers would prove economically and physically impractical. Just think of all those poor schoolchildren having to carry around textbooks weighing fifteen pounds apiece. (Kidding. Those kids didn't go to school; they picked through garbage heaps so books could be made for rich white guys.)

In order to cut costs and facilitate production, printers replaced the wooden slats with cartonnage (basically cardboard with a consistency closer to papier-mâché). But no gentrified man worth his weight in neck doilies would be caught dead carrying around a *cardboard* book, so binders covered the unsightly boards with a material as beautiful as it was sturdy: leather.

Today, most books are covered with decorative paper. About a hundred fifty years ago, cloth was the preferred material. For the hundreds of years before that, leather ruled the world of bookbinding.

As collectors of earlier imprints know, leathers from different sources result in different physical traits. One of the primary materials used in the early years of printing was vellum, made from the skins of calves, sheep, goats, or any convenient hide of high quality. Vellum tends not to be dyed, so these bindings most often appear white or cream colored. Some vellum bindings are so starkly white that they seem to glow like resplendent angels. Resplendent angels wearing something else's skin, but angels nonetheless.

When cowhide is tanned (instead of degreased), it's not called vellum; it's called calf, which is the most common leather binding. Now, if an eighteenth-century book owner wanted something really fancy, he'd choose morocco, made from goatskin. Morocco is a more expensive material, known for taking dye much better than other types of leather. Morocco bindings are truly stunning in their vibrant blues, greens, and reds. It's a pretty safe bet that if you see a shelf of leather-bound books in these hues, it's filled with morocco volumes.

Pigskin was occasionally used in certain locations, especially during the first few centuries of print. A blind-tooled alum-tawed German pigskin binding from the sixteenth century is a magnificent sight. That description probably means little outside the rare book world today, but it sure meant something to a sixteenth-century Jew looking to bind his copy of the Talmud. There have actually been cases when pigskin was used as the covering to books of Jewish scripture—which technically would have made their holy book no longer kosher. Unsurprisingly, European binders were more accommodating to their Gentile clientele, and at times that clientele had very specific tastes when it came to bindings. Books of erotica, for example, were sometimes bound in fur. Because of course they were.

De Caro's *Sidereus Nuncius* was not bound in fur (which would have been indicative of more "terrestrial" telescopic observations on Galileo's part); it was bound in vellum. Yet, baby farm animals weren't the only ones being used for bookbindings. Any animal that could be skinned has probably ended up around a book at one time or another. The skins of buffalo, snake, shark, salmon, frog, eel, and wallaby have all been used to bind books, as has ostrich shin (not skin, mind you, *shin*).

While rather uncommon, and obviously unethical, there are even examples of human skin being used as book leather. One American specimen comes from the *Narrative of the Life of James Allen*, a memoir of a nineteenth-century highwayman written while he was in prison awaiting execution. This armed robber had finally been brought to justice with the help of a man named John Fenno. Prior to his execution, Allen requested that his memoir be bound in his own skin and delivered to the man who had helped capture him. The book, completed in Boston in 1837, was indeed bound in Allen's skin, and delivered to and kept by the Fenno family for about seventy years before being donated to the Boston Athenaeum.

While owning a book made from human skin might justifiably make your own skin crawl, the more important question is: Who in God's name made the binding leather from James Allen? Who pulled his body out of the execution chamber and thought, *Well, it was his final wish, so I'll just go ahead and make a book out of him*? James Allen the highwayman was dead, but the monster who skinned him, dried out his hide, and then sewed it into a book was walking the streets of nineteenth-century Boston. Freely.

Thankfully for De Caro, vellum was a perfectly acceptable binding for Galileo's seventeenth-century anthology. He just had to ensure that his treatise fit seamlessly among the authentic sections. In order to accomplish this, he needed to address the edges of the book.

Historically, a book's edges have been a canvas for displaying wealth. Gilt decoration, in which gold is applied to the outer edges of the text block, immediately comes to mind. In many books, only the top edge will be gilt. This serves two purposes: to make it beautiful (very important) and to form an extra barrier to keep dust out of the pages. In order to trick out his reconstructed anthology (but, really, to cover up the scars of his Shelleyesque book surgery), De Caro inserted his forgery, cut five millimeters off the fore-edge of the entire text block, gilded the sucker, and pressed a decorative pattern into the gilt, a technique known as gauffering. This made it appear that both the authentic pages and the forged pages of the anthology had all been cut, gilded, and gauffered at the same time. Which they were. Just four hundred years later than the Italian book dealer was claiming.

De Caro chose to decorate the fore-edge of his text block with a spotted, abstract design that fit the era of the binding, but there are other, more elaborate ways a wealthy patron might display his status on a book's edges. An artist could be hired to create a one-of-a-kind fore-edge painting, for example. These strange works of art are made so that they're indiscernible when viewed directly from the front, but bend the block of pages in just the right way and—voilà! A secret painting appears!

Scenes of quotidian life, landscapes, religious motifs, and floral panoramas were common enough, but if someone shelled out the cash for an artist to hand-paint a concealed scene onto a book's fore-edge, what do you think a favorite subject would have been? If you guessed porn, then here's a gold star for you. Scenes of erotica were a popular subject for hidden paintings in books. Because nothing classes up your 1893 edition of *The Poetical Works of Henry Wadsworth Longfellow* like gentlemen and gentlewomen fornicating on the fore-edge.

From 2005 to 2012, the book dealer Richard Lan sent De Caro's *Sidereus Nuncius* to scholars and forensic investigators on two continents for inspection. De Caro's was perhaps the most successful book forgery of modern times, but he was still caught. Once doubt crept into the minds of the experts, it took only a few weeks for them to unmask a forgery that had taken more than two years to create.

Ultimately, it was small technical developments made throughout the hundreds of years of Western printing that undid De Caro's multimillion-dollar fake. In retrospect, a number of the fooled scholars were mortified to admit that there were signs everywhere. The depth of the pressed type, a faulty capital *P*, the sewing pattern on the binding, even the absence of wax on the binding thread became kindling for De Caro's eventual and inevitable immolation. Yet the book dealer's fall wasn't entirely a result of gaps in his knowledge of the history of print. Flawed people created that history, and in the end, De Caro proved to be De Caro's worst enemy.

Conmen like to produce a clever story to hide their real intentions. After the forgery was discovered, De Caro made the claim that he had "inserted a minor error or two," thinking that if he didn't, it would be impossible for history ever to uncover his masterful deceit. In a work of absolute perfection (which his *Sidereus Nuncius* was in fact not), De Caro asserted that minute marks of imperfection had to be added in order to balance out the universe. Like some kind of holy stigmata, a blemish would linger in the midst of perfection, reminding the common folk of the human origin of De Caro's work.

On the personal stamp of Federico Cesi, the storied owner of the Galileo anthology, De Caro and his alleged colleagues closed a tiny gap in the inner decorative border, one that is left open in every other authentic example of the Cesi bookplate. Because stigmata come in twos, he included a typo on the title page, changing the Latin word *periodis* (meaning "periods," as in "orbit times") to *pepiodis* (meaning jack-shit nothing because it's not a real word). De Caro claims that this was his personal finger in the eye of the academic community. He has also asserted that this taunting of academia's "mental pygmies" was the real motivation behind his forgery. This explanation is widely believed by the scholars involved in the book's examination. "It is our thesis," goes the definitive study debunking the 2005 *Sidereus Nuncius*, "that the [forgery] is a projected duel with the community of specialists."

This is likely dramatic overstatement by both parties. The explanation for the forgery is probably much duller: money. New technologies such as photopolymer plates (a type of plastic sheet made through photographic processes) can re-create the look of a printed page and are much cheaper and easier to obtain than an actual printing press and the necessary fonts. Because of this, and the potential money to be made, *someone*, eventually, was going to try to fool the world with a forged rare book.

Hubris lurks in the heart of every forger. Forgers must believe they are smarter than everyone else, more meticulous in their details than even the experts in their fields. If this weren't the case, they would never have the confidence to become . . . well, confidence men. Now that he's been caught, De Caro has tried to write the whole thing off as a joke, claiming that only a "real expert" could bring him down.

Unfortunately for him, a "real expert" happened to be working in Georgia at the time. Both the Cesi stamp and the "*pepiodis*" typo caught the eye of Nick Wilding, a historian of science at Georgia State University, and a specialist in Renaissance studies. Initially, senior scholars blew off Wilding's concerns. For example, the Cesi bookplate could be wrong without the entire printed treatise being forged. What were such quibbles next to one of the world's most respected art historians assuring the world that the watercolors of the moon had been created by Galileo himself? It was a fantasy come true . . . the key word here being *fantasy*.

Wilding persisted, however, and began a series of discussions with one of the leading scholars on the project, Paul Needham. Skeptical at first, Needham nevertheless listened, and decided that there were enough red flags to take another look at De Caro's copy. He took it to Columbia University and examined it next to an undisputed first edition. "With the two copies placed side by side, one feature after another . . . began to look suspicious, and within about twenty minutes Needham felt certain that [De Caro's copy] was a modern forgery." A flurry of emails followed between Wilding, Needham, Lan, and other experts throughout the United States and Europe. Once the excitement from one of the greatest finds in the history of science had worn off and the experts took a second look, this time primed for doubt, they easily broke the hollow façade of De Caro's forgery.

So it was that in 2014, the initial study that celebrated the discovery of this Galileo "proof copy" had to be supplemented with a volume that basically debunked everything that had come before. The correction reads like a confession from some of the world's most eminent book historians in the depths of serious soul searching: "not many results are worse than a refuted authentication." As Wilding himself has noted, labeling the forgery a "masterpiece" is one way the humbled scholars licked their wounds: in other words, only a masterpiece could have fooled them.

Humanity's capacity for brilliance is perhaps matched only by the depth of our egoism. Both materialize in the books we print—or, in the case of the *Sidereus Nuncius*, pretend to print. De Caro tried to pattern his life philosophy after that of Galileo Galilei. How this extended to international book forgery is a bit of an open question, but in the end, the two men did have at least one thing in common: their sentencing. Galileo, the Father of Modern Science, was sentenced to nine years' house arrest by the Roman Inquisition for heliocentric heresy. De Caro, not the Father of anything, was sentenced to seven years' house arrest by a much less notorious Italian court for being a thief.

TWO

2

FORGETTING MR. GOOSEFLESH

"WRITING BOOKS BY HAND MUST not be stopped because of printing!" Thus opens the seventh chapter of *De laude scriptorum manualium* (*In Praise of Scribes*) printed in 1492. That's right, *printed.* One of the most important critical arguments we have against the newfangled technology of the printing press was spread by the self-same technology it crusaded against. This is not entirely unlike a modern blogger who bemoans the encroaching digital age, slamming the Post button over and over again while pining for the good old days when people communicated using artisanal stationery and fountain pens. Okay, we might have done that once or twice. As it turns out, even after five hundred years, people haven't changed that much.

The author of *In Praise of Scribes* was a Benedictine monk named Trithemius. (That mouthful of a name is pronounced Tre-TAY-mee-us.) This monk's objections to the fledgling printing press, invented by Johannes Gutenberg about forty years before, focused on arguments such as the superior durability and craftsmanship of hand-copied manuscripts. "Who may not perceive how great are the differences between handwriting and printing?" Trithemius declared.

Each technological breakthrough hits resistance somewhere. Undoubtedly someone, somewhere, has said, "Automobile carriages can never match the dignity and maneuverability of the penny-farthing bicycle." Granted, that someone was your handlebar-mustachioed barista from last Tuesday, but historically speaking, technological advancement always picks up naysayers, and the printing press was no exception. Trithemius is the most infamous of those printing naysayers. Yet the main reason the world knows his name is because of the preservative power of print.

Consider, on the other hand, that the name Johannes Gutenberg appears in exactly zero printed books attributed to him. *Zero.* For almost three hundred years, there was no concrete evidence that Gutenberg was even the inventor of Western printing. In a grand twist of fate, the printing press preserved the life of one its greatest critics and all but erased the life of its creator.

Any schoolchild today can answer the question "Who invented the printing press?," but behind the answer so easily given lies centuries of scholarship, sometimes ingenious, sometimes not, in order to match Johannes Gutenberg with one of the most important technological innovations in human history.

WHAT DO we know of Gutenberg's birth? We know the place. He was born in Mainz, in what is now western Germany. As for his birthday . . . or birth year . . . or birth decade for that matter, we can only guess. Estimates place the date at sometime between 1396 and 1406. This means we don't know how old Gutenberg was when he invented the printing press, or even how old he was when he died. As for his death, we can only guess at those details as well. Given that this was the fifteenth century, we're just going to assume that it was painful and involved a shocking amount of bloodletting.

We do know his full name: Johannes Gensfleisch zur Laden zum Gutenberg, but even this behemoth took considerable effort to sort out. "Gensfleisch," "Laden," and "Gutenberg" refer to German households, with the Gensfleisch family owning the houses Laden and Gutenberg. This patrician system would fall apart by 1428, but by then it was too late, young Johnny Gutenberg already had an absurdly long last name. Even the name Johannes is not so simple in the historical record. Other common medieval derivations of "Johannes" that have been applied to Gutenberg include "Henchen," "Hengin," "Henne," "Henn," "Hans," and "Hansse."

Not understanding the intricacies of fifteenth-century German social structures, some later scholars thought Johannes G. zur L. zum Gutenberg was in fact two separate people: the brothers Johannes Gensfleisch (literally translated as "Jonathan Gooseflesh") and Johannes Gutenberg (or "Jonathan Goodmountain"). Of the many confused stories told about these two "brothers," it was said that not only were they responsible for the introduction of European printing, but they were also extremely shady dudes. The older brother was reportedly a thief who stole the secret of movable type from a master to whom he was apprenticed, and the younger brother was a heartless hound who broke off his engagement with a German maiden from Strasbourg. Amazingly, this dually libelous opinion of Gutenberg persisted even into the nineteenth century. It's one of those unfortunate quirks of history that for four hundred years, scholars knew very little about the inventor of the printing press, but they were pretty sure he was a scoundrel.

Within a generation of its invention, the details of the earliest printing press had fallen into confused obscurity. Some said printing was invented in Italy by the Frenchman Nicolas Jenson. Some said it was invented in Strasbourg, where Gutenberg lived for a time. Others suggested that the inventor was a man named Procopius Waldvogel, in Avignon, France. During Leipzig's bicentenary (1640) and tercentenary (1740) celebrations of the invention of print, the city praised Mainz businessman Johannes Fust and his partner Peter Schöffer as the inventors. Gutenberg was remembered only as their "assistant."

There simply wasn't enough early documentation to link Gutenberg indisputably to the greatest invention of the modern world. The earliest printed reference that unambiguously calls him the inventor didn't appear until 1472, almost two decades after the first printed books were sold. But with so many other candidates vying for the title, Gutenberg seemed no more attractive than anyone else. It wasn't until 1741, more than two hundred fifty years after Gutenberg's death, that a German history professor named David Köhler published a book entitled *The Vindication of Johann Gutenberg.* After decades of digging through archives in Germany, Köhler discovered several documents related to Gutenberg's life. In his pivotal work, Köhler printed a number of crucial manuscript records—at least one of which appeared to connect Gutenberg to the invention of the printing press.

Even with newly printed records, though, studies like these served only to thrust Gutenberg to the top of the contender's list, not to settle his claim once and for all. This is mostly because only two documents from his life reference printing explicitly—and one of those documents went missing for more than a hundred fifty years after its discovery in the eighteenth century. The documents people *could* find provided precious little detail. The story of who Gutenberg was and what he accomplished had to be inferred through the keyholes of a few surviving handwritten manuscripts.

Because the creator of Western printing didn't print anything about himself, we know almost nothing about Gutenberg's birth and early years. Trithemius, on the other hand, who made no secret of his criticisms about printing, wrote—and printed—extensively on his own life.

THE MAN who would be called Johannes Trithemius was born in 1462, six years before Gutenberg died, in the tiny village of Trittenheim, which sits about eighty miles from Mainz. The family into which Trithemius was born suffered from devastating poverty, and worse, Johannes's father died when his son was just a year old.

Being a single mother sucks, no matter what time period you live in. But being a single *Mütter* in fifteenth-century Germany, with no welfare, day care, or even women's education, had to have been near the top of the solo-parenting suck list. Johannes's mother, Elizabeth, walked the road of mother and widow for seven trying years before finally breaking down and remarrying. Why did she hold out so long? "[S]o that the tender babe would not be exposed, as was often the case, to the abuses of a stepfather." Elizabeth knew what every Grimm fairy tale teaches us: German stepparents are the absolute worst.

To exactly no one's surprise, it turned out Elizabeth was right. Her second marriage made her life more financially stable, but the cost was a father figure who hated learnin' like the pope hated Satan himself. Reading books was nonsense, and if young Johannes persisted in filling his head with letters and diagrams and whatever else was going on in those goddanged things, well, someone needed to pick up a leather lash and whip the persistence right out of him. But the constant physical abuse had the opposite effect on Trithemius, making him even more determined to study in secret and then get the hell out of Dodge (i.e., Trittenheim, Germany).

A particularly life-changing moment occurred when Johannes was fifteen years old. He had a dream that he would later interpret as a vision from God. In it, a young messenger approached him holding two stone tablets. On one tablet was written a collection of letters. On the other, a collection of drawings. Johannes was instructed to choose. Which do you like better? Words or pictures? Most people would have chosen the pictures, because most people back then couldn't read. But Johannes chose the words. Translated to a modern equivalent, it's like someone appearing in your dreams and asking you to choose between an article from *The New Yorker* and a viral video of a baby panda sneezing. You know you should pick *The New Yorker*, but this is an angel of God and it already knows you'd rather see that baby panda sneeze.

Trithemius, however, chose correctly. "Behold," the messenger announced, "God has heard your prayers and will grant you whatever you have asked." Within a couple of years, Trithemius officially declared his independence, waved good-bye to his stepfather with the appropriate, *Thanks for everything,* Arschgesicht, and openly pursued his studies at the renowned Heidelberg University.

IN CONTRAST to the numerous details of Trithemius's early life, Johannes Gutenberg doesn't even show up in the surviving historical record until sometime in his mid-thirties (probably). Between 1434 and 1444, his name appears in a few documents relating to taxes, loans, military eligibility, and lawsuits (especially lawsuits), but nothing that gives us a solid portrait of the inventor of the printing press.

For centuries we've been suing each other with obsessive fervor, and that constant barrage of litigation has created a steady fossil record of our civilization. Some of these cases have shaken the very foundations of society, such as the English *Case of Prohibitions* [1607] or the American *Brown v. Board of Education* [1954]. Others have provided some of the only clues to a certain fifteenth-century inventor's character and, without further context, have made him look like kind of an asshole. As Paul Needham points out, Gutenberg "referred to one of the [legal] witnesses against him, a shoemaker, as a sorry good-for-nothing who lived a miserable life of lies and deceit. One scholar who spent his entire professional life investigating Gutenberg questions wrote that this revealed the 'proud and temperamental' character of a Mainz patrician."

In 1436 a breach-of-promise suit was brought against Johannes Gutenberg by a Strasbourg woman named "Anne of the Iron Door." Gutenberg had allegedly entered into an agreement to marry this aristocrat, but reneged before the ceremony. Even though the exact outcome of the suit is unknown (except that Gutenberg never did marry Ms. Iron Door), his breach of promise provided the backbone for his being painted as a tomcat for centuries to come.

Whether we like it or not, lawsuits are miniature snapshots of society, and sometimes the surviving remnants are not conclusive, representative, or fair. Mark Twain poked fun at these kinds of spotty historical records with his "A.D. 5868" prediction of an encyclopedia entry for Civil War general and eighteenth president of the United States, Ulysses S. Grant:

> URIAH S. (or Z.) GRAUNT—popular poet of ancient times in the Aztec provinces of the United States of British America. Some authors say flourished about A.D. 742; but the learned Ah-ah Foo-foo states that he was a contemporary of Scharkspyre, the English poet, and flourished about A.D. 1328, some three centuries *after* the Trojan war instead of before it.

It's only too easy for us to read more into the historical record than is actually there. The realm of Gutenberg studies, with so much passionate devotion and so little surviving evidence, is notorious for this. Even the most celebrated document connecting Gutenberg to the invention of the printing press (a lawsuit, of course) has been misunderstood for centuries.

Twenty years after the breach-of-promise suit, Gutenberg was hauled into court again. We're not prone to utter this phrase often, but thank god for lawsuits. The outline of the evidence for and against Gutenberg, which would become the most important manuscript in the history of print, is called the Helmasperger Instrument. Helmasperger was the name of the signing notary, thus granting him the unimpressive title of history's most famous notary public. This 1455 suit addressed a sizable loan made by Johann Fust to Gutenberg to help cover the costs of his printing press. As with most historical documents, the parties involved weren't trying to explain details that were already obvious to them—such as how Gutenberg had invented the press, or how the press actually worked. The lack of detail is agonizing. Print historians today would trade every elbow-patched tweed blazer in the world for just one doodle in Gutenberg's hand.

Here's what we do know: between 1450 and 1455, Fust, a money broker from Mainz, lent Gutenberg 1,600 gulden. (A gulden was roughly equivalent to an ounce of silver.) Eight hundred of these gulden went toward the construction of a printing press. The other eight hundred were invested in various printing projects. For some reason, Fust wanted all this money back, plus interest, totaling 2,020 gulden. The court ruled that Gutenberg had to repay the printing press loan and anything else that wasn't considered a "joint project."

That's it. We don't know how much money Gutenberg was ordered to repay, or how capable he was of making payment. In fact, we don't even know the details of the printing press in question. The first surviving image of the printing press was published almost fifty years after this lawsuit, and it depicts a two-pull wooden screw press. (For some inexplicable reason, it also shows a bunch of zombies hanging around being jerks to everyone. Ah, the Middle Ages.) Gutenberg's press would have been a single-pull mechanism, and of such a press, "we know nearly nothing."

For centuries, the general consensus was that this lawsuit bankrupted Gutenberg. Other circumstantial evidence, such as his payments on a separate loan in Strasbourg suddenly ceasing around this time, appear to support this claim. After the suit concluded, Fust and one of the workers at Gutenberg's press, Peter Schöffer, opened a shop of their own. In just two years' time they had produced a stunning Psalter, the first printed book to contain a colophon.

This is a major historical moment. A colophon is a section at the end of a book where the publication information is printed, details that have since migrated to the title page or copyright page. The first colophon read, "[F]ashioned by a mechanical process of printing and producing characters, without use of a pen, and . . . completed . . . by Joachim Fust . . . and Peter Schöffer." It's incredible how much world-altering innovation lies just beneath the surface of that phrase "without use of a pen."

The 1457 Fust and Schöffer Psalter is a landmark for more reasons than the colophon. Their title page was printed in three colors: red, blue, and black. They also printed decorative initials and used two sizes of type for the first time. Schöffer had worked as a scribe before joining up with Gutenberg, and with his Psalter, it became clear that the physical beauty of books would play a major role in the history of print. Today only ten copies of that Psalter are known to have survived. A trip to the British Library to see one should be on everyone's bucket list.

Scholars have traditionally assumed Gutenberg must have been unable to pay his court settlement for two main reasons: Fust and Schöffer continued to print (gorgeous) books after the suit, and no printed works from that time forward claimed Gutenberg as their printer in the colophons. Thus scholars inferred that Gutenberg must have forfeited his printing press to his ex-business sponsor. Fust and Schöffer then cranked up their new press and produced a steady stream of revolutionary books and pamphlets, all prefaced with Nelson Muntz's catchphrase from *The Simpsons*: "Ha Ha!"

The cutthroat businessman taking advantage of the innocent inventor is a trope with a very long history indeed. Over the centuries, Fust was painted as a character so dastardly that he kept getting mistaken for Faust, the German scholar who, as legend has it, made a deal with Satan.

Scholars such as Paul Needham, however, have argued that Gutenberg kept printing, even after the 1455 lawsuit. It's likely that Gutenberg was responsible for the great Latin dictionary the *Catholicon*, first printed in 1460. We also know that, when he died, Gutenberg owned material for printing: a 1468 document describes one Dr. Humery inheriting "forms, letters, instruments, tools, and other things pertaining to the work of printing, which Johannes Gutenberg has left after his death." This means that the suit might not have impoverished him after all. In fact, it's possible that the Fust suit may actually have favored Gutenberg by allowing him to dismiss portions of the loan that were seen as "joint projects."

But this narrative was a long time coming. One scholar from the 1620s—who just happened to claim Fust as an ancestor—had used selective portions of the Helmasperger Instrument as proof that Johann Fust, with help from Peter Schöffer, was the true inventor of the printing press. According to this creative interpretation, Gutenberg, not Fust, was the wealthy moneylender from Mainz. In fact, Gutenberg was cast as a nosy neighbor who elbowed his way into one of the greatest inventions of Western history. *Oh, scheiße, here comes Gutenberg. Mein Gott, that guy is the worst. Pretend that you're not inventing a single-pull printing press.*

From Fust to random printers in Strasbourg, from Italian and Dutch inventors to thieves pilfering printing shops on Christmas Eve, the question of who invented the printing press has been a long and frustrating debate—and one that probably could have been avoided if Gutenberg had printed anything between 1450 and 1468 that read, *I did this! It was me! I'm actually a totally reasonable neighbor with original ideas of my own!* But, excruciatingly, we have nothing.

Yet there is always a tinge of excitement in the midst of this uncertainty. With so little evidence, a single new discovery could completely upend all we thought we knew about the inventor of the Western printing press. This isn't as crazy as it seems, either: a previously unknown eyewitness account of a Gutenberg Bible being sold in 1455 wasn't published until 1947. Even then, no one realized the depth of its implications for the history of print until a historian named Erich Meuthen translated and published the account along with an accompanying article in 1982. There are more stories like this than you might think: slightly earlier in the century, a fragment of a printed indulgence was discovered with the date filled out by hand, "22 October 1454," making it "the earliest precise date by which we know typographic printing was being carried out in Mainz." It speeds the pulse to think that more documents like this might be sitting around waiting to be discovered.

Ironically, because Gutenberg didn't attach his name to any of his productions, the main evidence we have of the inventor of print is a scattering of handwritten manuscripts. As we've seen, even the date for Gutenberg's now-famous Bible comes partially from a note nonchalantly scribbled onto one of its pages.

The most crushing problem with these manuscripts, however, is that the majority of the contemporary evidence referencing Gutenberg in any way has disappeared in its original form. Reading a summary of the known Gutenberg documents is like reading a literary casualty list:

> . . . original has not survived . . . last seen in Frankfurt in the mid-nineteenth century . . . original manuscript perished most

likely during the revolutionary upheavals at Strasbourg in 1793 . . . has not survived . . . last seen as late as 1830, but has not since been relocated . . . preserved in the old city library of Strasbourg and were lost when that library was destroyed by fire in 1870 . . . perished . . . not survived . . . lost . . .

WHERE GUTENBERG failed to be remembered, the printed words of an infamously eloquent critic of printing ensured that Trithemius would be. This Benedictine monk stood against the tide of technological innovation. In contrast to the disconnected, unfeeling clutches of the printing press, hand-copied manuscripts, he claimed, "give virtue to words, memory to things, and liveliness to times and circumstances." We're not sure what that's supposed to mean exactly, but at least Trithemius attributed the printing press correctly: "In those days in the city of Mainz, located in Germany on the banks of the Rhine . . . was invented and devised by the Mainz citizen Johannes Gutenberg that marvelous and hitherto unheard of art of printing."

Trithemius wrote this in his *Annals of Hirsau*, a fourteen-hundred-page history tome painstakingly written out by hand between 1509 and 1514 and eventually printed in 1690. Trithemius held a conflicted view of the "hitherto unheard of art of printing." Taken at face value, his *In Praise of Scribes* makes him appear as one of history's greatest antitechnology curmudgeons. There's Trithemius, some guy who thought "talkies" would never catch on, your aunt who couldn't figure out laptops in 2007, and everyone who's afraid that the Kindle marks the end of "real" books.

In that same antiprinting tract of 1492, however, Trithemius also wrote, "O blessed art of printing, long to be remembered as belonging to our age! . . . Now that this art has been discovered and you have been made our guide, it is henceforth permitted to any ordinary person to become as learned as he will." Now that's a statement we can get behind. In fact, Trithemius came to rely so heavily on one nearby printing shop that a recent biographer nicknamed it "the Sponheim Abbey Press." To understand how someone could be so for and yet so against printing, we need to dive a bit deeper into Trithemius's time at the aforementioned Sponheim Abbey.

After graduating from Heidelberg University, Trithemius and a traveling companion were returning home and stayed the night at St. Martin of Sponheim, a Benedictine monastery seventy-five miles from their destination. The next morning, their gracious hosts saw them off, but the two men were confronted with a sudden snowstorm in a mountain pass not too far from the abbey. Three separate times they tried to push forward, but the storm whipped up so violently that they eventually fell back to Sponheim. While his traveling companion left once the weather cleared, Trithemius interpreted the freak storm as the providential hand of God. The wind and snow and impending frostbite didn't say it outright—or anything at all—but the subtext was obvious to Trithemius: go back to the monastery, change careers, and become a monk for the rest of your life. Message 100 percent received.

Trithemius took his monastic vows when he was twenty-one years old. The next year, the head of St. Martin's had been transferred and the abbacy was offered to the young and ambitious Trithemius. For the next two and a half decades, he helmed the Sponheim Abbey. And life was good. There weren't any terrible stepfathers around, so Trithemius could spend his time studying and reading. Sure, there weren't a lot of books to read per se—besides a few copies of the Vulgate Bible, there were only eight volumes at the abbey when he first arrived—but as the new abbot, he could trade and purchase books to his heart's content—which he did, eventually accumulating a library of two thousand volumes, one of the most impressive and enviable of its day. Is there a better job on earth than creating a library? (That's a rhetorical question. We know the answer is no.)

Still, even the Garden of Eden had a problem with snakes. Trithemius's snakes were lazy, blockheaded monks who'd rather take the easy road than engage in God's true work: making books by hand. That's right, God wants you to make books. Using only your hands. Within the pages of *In Praise of Scribes,* Trithemius writes about the infuriating monks at the abbey, who made excuses such as (and this is in Trithemius's own words), "I cannot write, I am unable to remain in the cell for the entire day, I will gladly go to labors outside, I will not in the least refuse to dig or carry stones, but I pray only that I may not need to suffer distress in those things which are completely against my nature."

Oh, boo hoo. You can't sit in a dark cell and copy books by hand? Well, too bad! Life isn't supposed to be a picnic where you carry stones around all day! Trithemius wrote *In Praise of Scribes* in order to put his wimpy, whiny monks in their proper place—and this, forty-plus years after Gutenberg should have made hand-copying extinct.

For Trithemius, handwriting a text was a deeply spiritual, meditative act. There's a compelling point here. Writing cements the words in a copyist's mind. "[T]he printed work is a thing made of paper . . . but the writer who commends his writings to membranes [parchment or vellum] extends both himself and those things he writes far into the future."

Handwriting was a means of purifying one's soul toward Christ and preventing that most pernicious of monastic sins: idleness. By righteously keeping his hands busy in a dim cell each day, the Benedictine monk was said to reap the following rewards: "his most precious time [will be] employed fruitfully," "while he is writing, his intellect [will be] illuminated," his "affections [will be] enkindled on behalf of devotion," and "he will be endowed in the afterlife with a singular reward."

Then along came the printing press and screwed everything up. Now, sweaty, drunken craftsmen could do in hours what pious, meditative monks used to take weeks to accomplish. How is that the correct order of things? Serving God is supposed to be way harder than normal life. More gratifying, sure, but *harder.* Without copying books by hand, what would monks do all day? Lounge around on their luxurious rocks? No! By God, no!

The seventh and eighth chapters of *In Praise of Scribes* (out of sixteen total) do actually contain some notable criticisms about the emerging technology of print. "Handwriting placed on parchment [animal skin] will be able to endure a thousand years. But how long, forsooth, will printing last, which is dependent on paper? For if in its paper volume it lasts for two hundred years that is a long time." Forsooth, indeed. While this is a bit of an exaggeration, point taken: paper is a weaker material and, therefore, all else being equal, not likely to last as long as parchment. Yet paper's greatest vice is also its greatest virtue: relative to animal skin, it's cheap. Because books printed on paper were more cost-effective than parchment, they could be produced on a significantly larger scale. So, yes, a single book printed on paper might not last as long as a book handwritten on parchment, but the "Collective" of printed books has outlasted its animal skin counterparts at a rate of nearly two to one. *We are printed books. Your biological and technological distinctiveness will be added to our own. Resistance is futile.*

Very similar arguments have been made regarding the evolution from physical books to digital books in the twenty-first century. Trithemius would be appalled that fragile paper books would seem like mighty redwoods next to the immaterial ones and zeros of virtual space. You actually have to burn a physical book to destroy it, but a digital book can be obliterated with a high-powered magnet or a stumble over an electrical cord. It can even be remotely deleted against your will, as happened when Amazon erased illegal editions of George Orwell's *1984* and *Animal Farm* from every one of its Kindles in 2009. (The actual term for deleting digital books in this way is "dropping them into the memory hole"—a phrase ironically lifted right from the pages of *1984*.)

The poor availability of printed books in the fifteenth century was another point of criticism for Trithemius. Unlike manuscripts that had been in demand for three centuries, printing was still in its infancy. There were notably fewer printed books in circulation than hand-copied books. While this seemed like a big deal to Trithemius in 1492 (just like the limited availability of digital books might have seemed to someone in 2008), history has shown that after the initial start-up, the new technology will quickly surpass its predecessors. (Except for laserdiscs. And Betamax. And PDAs and Zip drives and QR codes. Okay, *most* innovative technology outpaces its predecessors.)

Not that the new technology necessarily kills the older ones. If print was supposed to kill manuscripts, it did a sadistically slow job of it. For hundreds of years after the invention of the printing press, works such as the poems of John Donne were produced specifically for circulation in manuscript form. Manuscripts didn't (and for that matter, printed books won't) suffer the quick and painless demise that many assumed.

There were copyists, Trithemius boldly declared, "who applied such a precise diligence to copying that they wrote not only correctly but also artfully, distinguishing their beautiful volumes by variation in notations, punctuation, and figures." In other words, manuscripts are prettier than print. Oh yes, there are some gorgeous hand-copied books out there. We fully agree on that. We're trying very hard not to begin rhapsodizing about the *Book of Kells*. But to use the aesthetic qualities of one technology to discount the value of another is where our roads fork.

The same argument could be made (and absolutely is) by people who mourn the transition to digital books. There's a different sensory quality to a printed book than a digital one: the feel of the pages, the sound of the spine cracking upon being opened, the smell of the paper. These are truly joyful—and addictive—pleasures of a printed book. Still, those sensations are not enough to wholly discount the virtues of the digital book. One of those unique virtues, for example, is discretion. No longer do you have to sit on a train, bending your book covers to prevent anyone from seeing your copy of *Twilight* or *The Da Vinci Code* or *Fifty Shades of Grey*. Now you can sit back and, with reckless abandon, enjoy some of the most embarrassing books humanity has to offer.

One literary scholar has summed up this knee-jerk resistance to technology by stating, "As we rationalize our resistance to digitalization, we reveal the fetishism of our relationship to the book. We are all too much like a toddler with a favorite, old blanket; the book comforts us because it feels good and we can carry it around." This is not to say that a physical book shouldn't be treasured, or isn't ineffably beautiful. A few favorites spring to mind right away. (We have a whole chapter on some of those.) But digital books can be beautiful in their way, too. There are coders, for example, who find algorithms "sexy." To quote Seinfeld, "Not that there's anything wrong with that." Sexiness, of the leather, paper, or digital variety, is, after all, in the eye of the beholder.

But of one thing we can be certain: resistance to technology is futile. For those who define themselves as against the tide, know that your struggle is centuries old. The objects to which you (and we) cling were once vilified by those who came before. Surely there's an ancient Egyptian out there somewhere turning up his nose at the thought of replacing his clean and elegant papyrus with processed animal carcasses. As the old saying goes, *plus ça change, plus c'est la même chose*—"the more things change, the more things stay the same."

Trithemius knew that his resistance to print was unpopular and, truthfully, impractical. Outside this conflict with the literarily lethargic monks of St. Martin's, Trithemius's regard for print was a "highly favorable one." Of the two thousand volumes in the Sponheim Abbey collection, twelve hundred were printed books. That means for every two spiritually endowed and lovingly created manuscripts in Trithemius's collection, he owned three that had been printed. He was able to create such an enormous medieval library because printed books made copies cheaper and more accessible. Very few book lovers would turn that deal down.

Thirteen years after he wrote *In Praise of Scribes*, the conflict with his monks became so contentious that Trithemius left St. Martin's and took up the abbacy of St. Jacob in Würzburg. It was the loss of his books from that move that affected him the most, the "library of two-thousand books which I assembled . . . and without which it sometimes seemed that I could not live." We feel for you, sir. Despite his tribulations, Trithemius found comfort in the printing press for its ability to restore at least a portion of his lost books. "The world today is abounding with volumes . . . For the art which they call printing . . . daily produces an almost countless number of texts by old and recent authors alike." It is human nature to hold on to your ideals with fists clenched, at least until the object of your indignation can benefit you personally. *Plus ça change, plus c'est la même chose.*

IT'S FAIRLY easy to track Trithemius, who wrote, and printed, dozens of treatises, from his "anti-printing" days at the Sponheim Abbey through to St. Jacob in Würzburg where he spent the remainder of his life. Gutenberg, on the other hand, is an entirely different can of *kerscheblotzer.* Gutenberg never wrote anything that survived. Not a description of his work, not a letter to a friend, not even a passing note throwing shade at Fust and Schöffer for the lawsuit. For hundreds of years we weren't even sure what the first printed book was.

In the three centuries after Gutenberg died, there was only one trustworthy near-contemporary description of the first book printed in Europe. In the *Cologne Chronicle* of 1499, the printer Ulrich Zell named Gutenberg the inventor of the printing press and claimed to have seen his first book: a Latin Bible with an abnormally large font. (A 1455 document, one discovered only in the twentieth century, described it as possible to read "without effort, and indeed without glasses.") So all someone had to do to discover the first book was to follow Zell's description and look through all the known books from the fifteenth century, setting aside any unattributed Bibles printed with unusually large type.

That should have been easy enough. How many books could Europe really print in fifty years? Well, it turns out about twenty-nine thousand editions have been documented. And that's twenty-nine thousand distinct editions on record. The number actually printed is probably much higher: recent scholarship suggests, not without significant pulling of hair and gnashing of teeth, that around 30 percent of editions from this period have been lost entirely. Even taking only the number of documented editions and using a very conservative estimate for each print run (say, around one hundred fifty individual books per edition), you're looking at a number of single volumes that reaches well into the millions. It's kind of like someone from 2041 asking, "C'mon, how many Web pages could realistically have been created in the first fifty years of the Internet?"

Books printed before 1501, within the first fifty years(ish) of Gutenberg's invention, are called "incunabula." This comes from a Latin term coined in 1640 meaning "in the cradle." In other words, the babies of print. Incunabula are essentially the only books considered rare and collectible simply because of their age. A book from 1830 might seem old from our modern perspective, but in the collecting world, it's not. Books have been printed for hundreds of years, and the nineteenth century was hardly that long ago in comparison.

Incunabula (and its singular, *incunabulum*) is a bit difficult to say, so these days many specialists refers to such books in conversation as "incunables." If you think that doesn't sound quite fancy enough, you're not alone. This Anglicizing of Latin is relatively new, and has been frowned upon by the more snobbish of the rare book community. Here's a delightful statement from a book on collecting published in 1982: "The Englished forms 'incunable' and 'incunables,' though a century old, are to be spat at." I guess we all have our silly lines to draw in the sand. For example, it is an irrefutable fact that the term *fifteeners*, another modern coinage for *incunabula*, is an inexcusable abomination before all that is holy and good in this world. That's just a fact.

Naturally, some of our earliest sources about the first printed book are incunables themselves, and they can be spotty, garbled, and contradictory. The aforementioned Ulrich Zell had been printing since at least the early 1460s, and was probably trained in Mainz. As far as patchy historical records go, his mention of Gutenberg's Bible was one of the firmest leads a historian could ask for. Scholars delved into their stacks of surviving printed books, triangulating requirements: Bible, large type, no printer information. Eventually, in the 1760s, the thousands of incunabula candidates were reduced to just two. Two unattributed fifteenth-century Latin Bibles with large type, each seemingly excellent candidates for the title of first printed book. One of these Bibles was nicknamed B42 (the font size allowed forty-two lines of text to be printed on each page), and the other was B36 (obviously for a similar reason).

With these two Bibles in hand, it still took the world's best print scholars more than a century to determine which had come first. Not until 1890 did Karl Dziatzko demonstrate in a cool bit of biblio-sleuthing that certain textual elements in B36 could have come only from a compositor who made copyediting errors while using B42. This definitively proved that B42 was older, and the rightful heir to the title: the Gutenberg Bible.

To think, until the end of the nineteenth century, scholars (or any Johann Doe, really) could have been reading the B42 Bible completely unaware that he was holding one of the first books ever printed in the Western world. Yet even if a reader never realized the importance of the book in front of him, he still would have been struck by the absolute beauty of it.

The first time you see a Gutenberg Bible is a spellbinding experience. Yes, you know you're going to be impressed because you know how historically important it is. That doesn't prepare you, however, for the impact of actually standing in front of the thing. This is one of those moments in life you'll always remember.

Why is Gutenberg's Bible so beautiful? Part of the answer can be traced to the competition he was facing. He had to compete with a manuscript book market that was already well established and thriving. If his new product were to succeed, it had to spring into life fully formed and flawless, like Athena from the head of Zeus. Gutenberg was trying to cut into the robust market of handwritten books, many of which were richly decorated and illuminated in an aesthetic tradition developed over a thousand years.

The Gutenberg Bible was printed as a double folio, fairly large by today's standards—each page measured roughly a foot wide and a foot and a half tall. It's partially the book's size that first strikes the viewer, conveying a sense of dignity and splendor. That size wasn't just for decoration, though. Gutenberg Bibles were characteristically large because they were meant to be used as pulpit or refectory books, practical volumes "that could be produced relatively inexpensively and purchased without great sacrifice by churches or monasteries." Gutenberg isn't often given credit as a savvy entrepreneur, but he knew his audience, and he created the perfect product for it. Monasteries were one of the largest consumers of books, and thus, large-format Bibles were the safer bet. Many of the B42 Bibles that exist today have a long history of ownership (called provenance) within monasteries, attesting to Gutenberg's success in targeting that demographic.

Current estimates suggest that around 180 copies of the Gutenberg Bible were printed, a third of which were printed on vellum—which is impressive when you consider that there are about 1,300 pages in these Bibles. At four printed pages per folio-size sheet, and one animal yielding between two and four sheets, each single copy of a vellum Gutenberg Bible would have required a whopping 170 calves or 300 sheep. For a modern comparison, consider that the largest herd of cattle in the United States exists on a ranch spread out over three counties in central Florida. Deseret Ranch, owned by the Mormon Church, counts 44,000 head of cattle—or roughly enough vellum to make only 250 Gutenberg Bibles. So a tip of the hat to paper. You rock. Europe's cattle industry never could have supported book production on the scale necessitated by the printing press. But paper could.

Public sales of the Gutenberg Bible have become some of the biggest events in bookselling history. One of the most famous of these sales occurred in London in 1847, resulting in the first Gutenberg Bible brought to the United States, entirely by mistake. James Lenox, a millionaire from New York, had authorized an agent to bid on his behalf at the March 12, 1847, sale of the library of one James Wilks. Because sending letters back and forth across the Atlantic Ocean would have made the auction last *forever*, Lenox's agent, the awkwardly named David Davidson, had to use his best judgment when placing bids. It turns out that Davey-squared got a bit excited. It was an auction. It was the Gutenberg Bible. Who are we to judge? This agent fell into a bidding war with the famed bibliophile Sir Thomas Phillipps and the result was not cheap.

Before the Wilks sale, the record price for a Gutenberg Bible was £215. Sir Thomas planned to bid up to an astounding £300. When Davidson outbid him, Sir Thomas was not pleased. Caught in the urgency of the moment, Sir Thomas started placing bids way beyond his carefully planned limit. Between the two, the price just kept galloping toward the horizon like some gorgeous, financially unconcerned unicorn. The bids ping-ponged back and forth until Sir Thomas's agent finally stepped in to "arrest [Sir Thomas's] mad career" at £495. Davidson placed the winning bid at a mind-boggling £500, more than doubling the previous Gutenberg record. This "mad price" became a sensation on both sides of the Atlantic.

When Lenox discovered that he was the lucky owner of America's first Gutenberg Bible, he was not thrilled. Okay, that's putting it lightly. Arguing that he hadn't authorized such an unreasonable and exorbitant price, he initially refused to pay it at all. His agents were left with an enormously expensive book on their hands. They approached the aforementioned Sir Thomas (no doubt sheepishly) and asked if the gentleman might be interested in purchasing the Bible for £475 (a 5 percent discount!). Sir Thomas wrote the following reply, "Sir, as I would not give more than 300 pounds for [the Bible] in cold blood, there is no chance of my having the Book, and it is right that America should boast of having one copy of it. I am therefore willing to let it go, only hoping that it may not be swallowed up by the Deep Sea." After the public embarrassment at the auction house, Sir Thomas must have felt awfully good about sending that letter. Lenox eventually did recover from the sticker shock and made the reasonable decision to accept the Gutenberg Bible at the unreasonable price of £500. If one of these Bibles appeared on the market today, it would easily fetch tens of millions of dollars.

Yet that's unlikely. Most modern owners of Gutenberg Bibles are revered institutions that would never trade a cultural icon of human achievement for a measly thirty million dollars. Then again, a single Gutenberg Bible was put up for sale in 2009, albeit to a very select group of consumers. The extraordinary circumstances of that case date back to the Second World War.

In May 1945, just after the fall of Nazi Germany, triumphant Soviet forces entered the city of Leipzig and looted it, secretly carrying away two Gutenberg Bibles. Because the Soviet Union was run by cultural sociopaths, they denied the possession of these Bibles well into the 1990s. It's not like the Russians gave them back, even after admitting the theft. In their view, the Motherland had defended itself against the Nazis, so Germany could, with all due respect, fuck off. Then, in an ironic twist, one of those Gutenberg Bibles disappeared from Moscow State University.

In 2009, three agents working for the Federal Security Service, the modern successor to the KGB, took advantage of their positions and stole one of the stolen Leipzig Bibles. The Bible was eventually recovered in a sting operation set up by the FSS itself after the conspirators attempted to sell the Bible for the shockingly low price of one million dollars. "These people were not art specialists," commented the court spokeswoman.

The leader of this idiotic band, one Colonel Vedishchev, was sentenced to more than three years' penal servitude. The reclaimed B42 reportedly had to undergo "repair work after a page was cut out for the buyer to check its authenticity." That one sentence is enough to cause a seizure in any rare book dealer, but at least we've learned that testing the authenticity of a black market book is essentially the same process as testing the quality of a kilo of cocaine.

From the very beginning, it seems the question of who owned Gutenberg's creations has been disputed. Whatever the verdict of the infamous lawsuit brought against Gutenberg by his ex-partner

Johann Fust, current scholars believe that after B42 was completed, the joint "work of books" between the two men officially ceased. Frustratingly, the little we do know about Gutenberg's work serves to highlight just how much we don't. For example, we have no idea where Gutenberg's print shop stood. Because the Fust shop was in Mainz, it seems reasonable that Gutenberg was also printing there, but without a single document containing his business address, it's hard to say where he was printing, or even if he closed up shop after the lawsuit.

The picture of Gutenberg's financial endeavors is complicated even further by the recent (if you consider 1889 "recent") theory that Gutenberg was actually running two separate print shops in the 1450s. One shop would have been printing B42 in association with Fust and Schöffer. The other would have completed "jobbing printing," that is, fast printing jobs that resulted in immediate cash. You know, your typical moneymakers like indulgences for the Church and calendars for bloodletting. These flotsam-and-jetsam printing contracts are some of the earliest known specimens of print, but many of today's known fragments survive only because they were used as worthless scrap paper in other books. In the nineteenth century, for example, a printed bloodletting calendar from the 1450s was found inside the binding of an accountant's book. Yep, that seems about right.

To recap: we don't know when Gutenberg was born; we don't know where he worked; two-thirds of the original manuscripts mentioning him are no longer extant; even the church where he was likely buried was demolished three hundred years ago; and we have absolutely nothing printed by him that contains his name. Enough is enough! All this could have been avoided if Gutenberg had just printed his motherfucking name on his motherfucking books. (The authors apologize for using expletives in connection with the Gutenberg Bible. It's a rather accurate representation of our scholarly frustration. Also, if Samuel L. Jackson had been a Gutenberg historian, it's what he would have said.)

While unearthing the first printed book in Western history and assigning it correctly to Johannes Gutenberg has been a Herculean achievement, other books and authors have been memorialized by print despite their best efforts to the contrary. And this is how Trithemius, a Benedictine monk from a small abbey in Germany, came to be remembered as one of the foremost dark magicians of the Catholic Church.

SCHOLARS GENERALLY organize Trithemius's writings into three main categories: monastic issues, histories, and demonology. We're hoping that last one made you pause. Trithemius was into some strange stuff, and he was eventually outed in a most unfortunate way. In 1499 he wrote a letter to a friend and fellow monk, Arnold Bostius. The two monks had the kind of relationship where Trithemius felt comfortable opening up about his views regarding the true powers of the universe, and man's ability to control those powers through, well, magical means.

These days, incantations and pictographs aren't normally associated with Benedictine monks. They weren't in Trithemius's day, either. If it had gotten out that Trithemius was deeply interested in the conjuration or banishment of demons, angels, and witches, it would have created a small scandal. So Trithemius kept his views close to his chest and discussed them only in the privacy of selected personal letters, such as the ones to Arnold Bostius.

Here's a life tip, dear reader, directly from Trithemius to you. If you want to keep a secret *secret*, never write it down. Never. Not anywhere. Not anytime. It's like when you're at work and you email your friend across the room about Susan being a total bitch. The thing is, Susan is going to read that email. She will. If you write it down, chances are she (or someone sympathetic to her) is going to see it. Which is fine, if you don't mind snuffing out the last defense mechanism Susan held on to after the divorce. You monster.

Trithemius also learned this lesson a little too late. When he sent his letter to Bostius, he had no idea that the intended recipient had died just a short time before. The ill-fated correspondence then found its way into the hands of the prior of the convent, a man who wasted little love on the outspoken Trithemius of Sponheim Abbey. Rather than quietly slipping the letter into a rubbish bin, the prior did exactly the opposite and circulated it as a public denouncement of Trithemius, practitioner of black magic.

What did Trithemius write to Bostius that got the prior's habit in such a knot? He informed his friend that he was furiously working on a series of books addressing the very important subject of steganography, which sounds innocent enough. Today steganography is a computer coding term for concealing a text or image within another text or image. But the person who invented the word had something a bit more unusual in mind. Derived from the Greek words *steganos* ("concealed") and *graphein* ("to write"), steganography, to Trithemius, was a framework by which human beings could transmit messages over great distances with total fidelity, using an occult system of angel conjuring. It was like a fifteenth-century telegraph, only powered by miracles instead of science.

Trithemius told Bostius that a shadowy figure had appeared to him in the middle of the night and taught him the ancient secrets of steganography. These secrets were summarized in the letter, and Trithemius informed his friend that he was working on something called *Steganographia*, a four-book instruction manual on this arcane method of secure communication. The fallout from his public outing, however, forced Trithemius to abandon the project partway through book three.

Some people reading this might not find Trithemius's idea so odd. Isn't transmitting messages the traditional job of angels? The Greek word *angelos* means "messenger," after all. Well, Trithemius's angels were less "hark the herald" and more Book of Revelation. If incantations were said wrong, or symbols drawn too hastily, or the wrong direction was faced, these angels could wreak righteous havoc on their summoners. Steganography, Trithemius wrote, "is very difficult and full of danger because of the pride and rebellion of its spirits, who do not obey a man unless he is very experienced in this art . . . if they should be too strongly pressed on, they frequently do injury to [the novice] and offend them by various delusions."

Other dangers one might encounter when trying to control rebellious spirits include angels bursting in on recipients at inappropriate times; spirits working themselves into an unholy frenzy and alerting everyone in the village; and giving your letter to the one person from whom you wish to keep it secret. *Aw, c'mon regional angel Pamersyel, you handed that email right to Susan!*

Recently there have been suggestions that Trithemius's *Steganographia* was a clever ruse to present his studies on cryptography within the guise of cabalist magic, but to say that trivializes Trithemius's decades of private arcane beliefs and writings (with titles ranging from *Antipalus maleficiorum* and *De demonibus*, to *De morbo caduco et maleficiis*). Cryptography certainly plays an important role in *Steganographia*, but if you were going to hide your involvement in cryptographic studies, why choose a framework that would blacklist you from the religious community and forever brand you a dark magician? In 1508, Trithemius completed a "natural" version of his work entitled *Polygraphia*. It was like *Steganographia*, but without the angel conjuring—essentially just a simple handbook of codes and ciphers. (Note that "simple" refers only to angel conjuring: the *Polygraphia* is actually a brilliant landmark of codicology known for its meticulous descriptions. The first edition now sells for thousands of dollars and is worth every penny.)

Steganographia certainly wasn't something Trithemius wanted out in the public eye—and with good reason. Almost a century after his death, Trithemius's *Steganographia* was printed in Frankfurt. Three years later, the *Index Librorum Prohibitorum* of 1609, the Catholic Church's official list of banned books, branded it a work of heresy. As long as manuscript copies were being passed from town to town by a quiet network of occult supporters, Trithemius's unique mix of arcana, Christianity, and spirit summoning posed little threat. But hand that sucker over to the long arm of the printing press, and Rome itself sits up to take notice. *Steganographia* remained on the *Index* for the next three hundred years. Thanks to the printing press, Trithemius's demonology, "a minor current" when compared to the rest of his writings, became his most definitive historical trait. Maybe he was right to hate Gutenberg's invention after all.

Trithemius died at St. Jacob's in Würzburg in December 1516, almost fifty years after the death of Gutenberg. A champion of educating the clergy, he served as an abbot from age twenty-two until his passing at age fifty-four. During that time, he built libraries, authored a flood of religious and historical treatises, and played host to princes, dukes, and emperors. Yet a pamphlet aimed at shaming his lazy monks and a treatment of arcane cryptography that was never supposed to be published have enshrined him within the halls of history as the magical monk who hated printing.

Still, at least Trithemius was remembered, due in no small part to the printing press. For hundreds of years, Gutenberg was not remembered at all because he never printed his own name. One of the most important technological innovations of human history almost passed through the annals of time disembodied from its real inventor. Had it not been for a scattering of handwritten manuscripts that were printed in the 1700s, Johannes Gutenberg would have been sentenced to obscurity. Whether on papyrus, vellum, paper, or digital text, sometimes the wrong people are forgotten, and sometimes those who are remembered are remembered for the wrong reasons. *Plus ça change, plus c'est la même chose.*

CHAPTER THREE

3

TREES OF TRUTH

THE REIGN OF KING HENRY VIII during the first half of the sixteenth century indelibly changed the course of English history. And one of the stupidest fights during that reign began in 1511, just two years after Henry was crowned. In that year a boy was born to a tailor and then promptly died five weeks later. Of itself this isn't noteworthy, especially during a time when living to thirty was a cosmic crapshoot, but the events that played out after the funerary services sent shock waves throughout the city of London, and epitomized the need for serious religious reform.

The word *mortuary* meant something very different in the sixteenth century than it does today. When a grieving father, Richard Hunne, took his son's body to St. Mary's Church to be buried, the priest asked for a mortuary after the services were completed. In this context, a mortuary is a gift to the officiating priest from the deceased person's estate—a death tax, of sorts. But because the deceased was a five-week-old, he didn't have much property to tax. In fact, it could be argued that a dead infant doesn't possess any property at all, so he should be exempted from the normal ecclesiastical payoffs. Yet thinking *that* would make you quite a bit more compassionate than Richard Hunne's priest.

In May 1512, this priest sued Richard Hunne in a London Bishop's Court, a spiritual court ultimately controlled by the pope, for the sheet in which the dead child had previously been christened. And if that weren't crazy enough, he won the case. The soon-to-be Bishop of London, Cuthbert Tunstall, ruled in favor of the priest and demanded that Hunne surrender his dead son's sheet. What on earth would a priest even do with a young boy's sheet? In the twenty-first century, it would probably be best to immediately abandon this line of questioning, which is fine, because the intentions of the priest are beside the point. Richard Hunne had questioned the actions of the Church. Therefore, he had to be put in his place.

Hunne stood his ground. He flat-out refused to hand over the sheet. As a result, he was publicly denounced by the Church: "Hunne, thou art accursed." That's right, for not relinquishing the one memento of his dead son, Richard Hunne saw his soul damned to hell for all eternity. In response, he brought a praemunire action against the priest, a provision of the law that asserted the authority of the king and his temporal courts over the pope and his spiritual ones.

Before the praemunire suit was settled, however, a raid on Hunne's house turned up some religious books in English. The possession of said books had been outlawed more than a century earlier. The most powerful charge against him was that he "hath in his keeping diverse English books prohibit and damned by the law: as the Apocalypse in English, Epistles and Gospels in English, Wycliffe's damnable works, other books containing infinite errors in the which he hath been long time accustomed to read, teach, and study daily." On December 2, 1514, Hunne was formally charged with heresy, arrested, and interred in a tower in St. Paul's Cathedral. Two days later, the London tailor was found hanging by the neck from the rafters in his cell.

The Church and its defenders, people such as Bishop Tunstall and the fiercely orthodox statesman Thomas More (more on this fellow later), declared Hunne's death a suicide. Case closed. Move on, people. Then details of the case began to emerge. Things such as Hunne's wrists showing evidence of being bound, the stool he used being found clear across the room—oh, and then there were hand marks on his neck, where he manually strangled himself before the hanging. The inquest investigating Hunne's death noted these details and more: "Also we find in a corner, somewhat beyond the place where he did hang, a great parcel of blood. Also within the flap of the left side of his jacket, we find a great cluster of blood, and the jacket folden down thereupon; which thing the said Hunne could never fold nor do after he was hanged. Whereby it appeareth plainly to us all, that the neck of Hunne was broken, and the great plenty of blood was shed before he was hanged."

All this notwithstanding, the Church stuck to its guns. In fact, it doubled down and declared Richard Hunne's *corpse* a heretic. The Church tried the dead tailor's body on December 16, found it guilty, and burned it at the stake. This was done either as punishment for heresy, or to prevent a potentially damning murder witness should zombies be real.

Scandals like this strongly turned the tide of public opinion against the Catholic Church and its clergy. For more than a century, reformers had been decrying the kinds of depravities that made the Hunne case so sensational: mortuaries, indulgences, legal favors granted to the Church, disproportionately brutal punishments, large-scale attempts to exercise control over every aspect of a person's life from cradle to grave. Religious reformers had come before the sixteenth century, but they all lacked that certain *je ne sais quoi* to make their regional movements global. Enter the printing press. Curtain up on the Reformation.

BEFORE MARTIN Luther nailed his Ninety-Five Theses to the door of the All Saints' Church in Wittenberg,

Germany, the printing press was not an ally of religious reformers. During the eighty years since Gutenberg's press was invented, the Church had used it to print indulgences. These were, in fact, perhaps the first thing Gutenberg ever printed. Indulgences allowed sinners to pay cash in order to avoid severe penances, and in some cases guaranteed salvation from eternal hellfire. One couplet from this period goes, "As soon as a coin in the coffer rings / the soul from purgatory springs." If purgatory was shooting out sinners like clowns from a circus cannon, Gutenberg's press was printing off tickets for the ride.

This did not sit well with one German friar and theologian. Indulgences, simony (selling church offices), and an unholy obsession with holy relics were at the top of Martin Luther's complaint list. This last one had been reaching plainly absurd levels for hundreds of years. The famous Dutch humanist Erasmus pointed out that there were enough wood splinters from Christ's True Cross to fill a ship. One ballsy bishop from Lincolnshire traveled to Fécamp Abbey in France, and when he was shown its holy relic, purportedly the arm of Mary Magdalene, he promptly bit off a couple of fingers. This wasn't a quick job, either. The bishop found it too difficult to snap the cadaveric digits with his front teeth, so he gnawed his way through them with his molars, much to the shock and indignation of his theretofore gracious hosts. For this act of cannibalistic theft, the bishop was showered with praise back in Lincolnshire, where he returned with two severed bits of St. Mary's fingers. He was consequently canonized as a saint himself in 1220.

Relatively benign practices pushed to their extreme, corrupt, and often ridiculous ends prompted Martin Luther to write a ninety-five-point protest to his bishop and, according to then-current tradition, nail it to a church door. This list of objections in Latin was originally intended for a narrow audience—"a few of my neighbors." Church doors were a bit like local bulletin boards for priests, circulating news or information such as the announcement of a lost kitten or objections to the supreme authority of the Catholic Church. Martin Luther's Ninety-Five Theses was a first draft of his complaints, submitted to a close circle of academics. "My purpose was not to publish them," he wrote to one printer. "Thus I [thought to] either destroy them if condemned or edit them with the approbation of others. But now that they are printed and circulated far beyond my expectation, I feel anxious about what they may bring forth."

Feeling anxious was right. The insatiable machinations of the printing press were about to spread Martin Luther's words throughout all of Europe and launch a religious revolution in his name. Luther himself thought the presses a difficult beast to control. "This method is not the best adapted to instruct the public," he mused, "[I] should have spoken far differently and more distinctly had I known what was going to happen." So he sensed in 1518 what most reality show stars in the twenty-first century learn only *after* they've humiliated themselves on national television: I should have been more careful about the shit I was saying.

Once Luther learned to embrace the printing press, his ideas spread farther and faster than he ever imagined. Luther expected a rational and scholarly discourse to follow, but Church authorities saw his move as inescapably treacherous. Publishing his first treatise in German (instead of Latin) was a major turning point in that dispute. Some historians believe the true beginning of the Reformation happened the moment Luther started printing in the common German tongue. His campaign enabled the average citizen to read (or at least hear) arguments about religion in a way he or she could easily understand.

To the Church authorities, already reeling from accusations of corruption and extortion, Luther's printing campaign exploded onto the scene like a flurry of multinational body blows. The criticisms became so widespread, in fact, that these popular Reformation pamphlets were later dubbed *Flugschriften* ("flying writings"). There was no turning back. As print historian Elizabeth Eisenstein famously said, "Heralded on all sides as a 'peaceful art,' Gutenberg's invention probably contributed more to destroying Christian concord and inflaming religious warfare than any of the so-called arts of war ever did."

FIVE YEARS before Luther wrote his Ninety-Five Theses, a man named William Tyndale graduated from Oxford University. He was a gifted linguist, but when he went to school, he found it stifling and, at times, ludicrous. Rather than enlightening the common man on the finer points of the Bible, theologians were spending their time debating issues such as how many angels could fit on the head of a pin, or if Christ had turned himself into a mule, could that mule have been crucified for the sins of the world. All cogent points, but Tyndale cared nothing for acrobatic angels or mulish sacrifices. He cared about one thing: an English-language Bible. In early sixteenth-century England, however, that was heresy.

Up to this point, the Bible had been largely protected from the gaze of working-class yokels by that most nefarious of locks: Latin. Literacy itself would have been a huge barrier, but double padlock that sucker with a dead language and you've virtually guaranteed that peons won't read it. Most Christians in the Middle Ages had no way of knowing for themselves what was actually written in the Bible.

Unfortunately, this same lock guaranteed that some of the clergy wouldn't read the Bible, either. One survey from 1551 found that out of 311 clergy members questioned, "9 did not know how many Commandments there were, 33 did not know where they could be found in the Bible, and 168 could not repeat them." If this seems impossibly absurd now, thank William Tyndale, who made it his life's work to open the contents of the Bible to everyone. Tyndale reportedly told one priest, "If God spare my life, ere many years I will cause a boy that driveth the plough shall know more scripture than thou dost."

The last thing the Catholic Church wanted was an "Arthur Cobbler" or "Hans Hoe" (both actual "Joe the Plumber" names for common folk) poking around in the founding documents of its religion. After all, the Hoes and Cobblers and Plumbers of the world were either too stupid or too wrapped up in their farms and shoes and online cat videos to really be able to comprehend the words of God. They were bound to misinterpret the Bible, and those erroneous views would lead to chaos. So if literacy and Latin weren't big enough barriers, the Catholic Church in England went ahead and made it a crime punishable by death to translate, distribute, or even possess an English translation of the Bible.

"It was impossible to establish the lay people in any truth," Tyndale wrote, "except the scripture were plainly laid before their eyes in their mother tongue, and they might see the process, order, and the meaning of the text." But by making it morally and legally forbidden to translate the Bible into English, Tyndale accused the Church of making Christ "not the light of the world, but its darkness."

After a brief stint as a family chaplain in Gloucestershire, William Tyndale decided to follow in the footsteps of revolutionaries such as Martin Luther and translate the Bible into his mother tongue. One Catholic supporter saw the danger of this and would later comment, "Even the tailors and shoemakers, and indeed women and simple idiots [now] read it."

Yet, for laypeople, reading the Scripture in their own language was important enough that "it was said that a man would give a cartload of hay for a few sheets of St. Paul." Robert Benet, a wool racker and water carrier, sold his looms and shears to purchase a copy of the English New Testament. Considering that the possession of said looms and shears is pretty much what qualifies you as a wool racker, that was huge. Thankfully Benet had carrying water around in buckets as a fallback career. It didn't even matter that he wasn't literate enough to read his English New Testament. He just carried it around with him in his belt, a symbol of his faith and the importance of the Word of God.

Because a theology degree doesn't pay the bills (unless you're open to kickbacks on infant funerals), Tyndale needed to find a wealthy patron to fund his Bible translation. This brilliant linguist surveyed England and homed in on a single name: Cuthbert Tunstall. Sound familiar? It should. He's the guy who ordered Richard Hunne to give up his dead son's sheet. Yet this is something important to note about Tyndale from the start: he was a moron. No, no, that's a bit harsh. Tyndale was an ingenious translator—in addition to English, he could fluently translate Latin, German, Italian, French, Greek, and Hebrew—a cunning and slippery fugitive, and a master of rogue printing and international smuggling. Politically speaking, though, he struggled to rub two brain cells together. This will be an ongoing theme with him, and it started with his terribly dim idea to petition Cuthbert Tunstall, now the Bishop of London, to sponsor him while he committed the felony of translating the Bible into English.

Fortunately for Tyndale, Tunstall replied with a curt but polite *no, thanks*. As Tyndale recalled, "My lord answered me his house was full . . . and advised me to seek in London." That's nice. This seems like one of the few times you're glad the guy in charge doesn't take your ideas seriously. Especially considering that three years later, this bishop would preach a public sermon against Tyndale at St. Paul's Cathedral and then oversee a mass burning of his New Testaments. (Ironically, the book burnings didn't have the effect

Tunstall had hoped for. Orthodox agents began buying up copies to be burnt, which only upped the demand and encouraged printers to issue more pirated editions.) Two years after the burning, Tunstall would launch a heretic witch hunt and interrogate people caught with Tyndale's translation in the very chapel where Tyndale had petitioned for help.

In other countries (notably Italy and Germany), vernacular translations of the Bible were widely available within a generation of the printing press. But England had an unfortunate run-in with Bible translations in the fourteenth century, and the country wasn't about to legalize a book that had fomented religious revolution.

While partial translations date back to the seventh century, some nine hundred years before, the version that ruined it for the Brits was produced in the late 1300s under the direction of theologian and university professor John Wycliffe. This professor started a reformation movement that would be nicknamed the Lollards. It is believed that *Lollard* comes from the Dutch word, *lollen*, which means "to mumble," and was "used for any kind of vagabond or religious eccentric." As with the Shakers, Quakers, and Mormons, the nickname was originally meant as derogatory, but people eventually forgot the joke, and adopted it.

What were Wycliffe's followers all mumbling about? The English Bible, translated and painstakingly hand-copied for the use of the common man. The Lollards were known to value English-language Scripture so much that any translation automatically became associated with Lollard heresy. Common-tongue translations were officially outlawed at the 1407–9 synod at Oxford led by Archbishop of Canterbury Thomas Arundel. (This archbishop, incidentally, made it onto *BBC History Magazine*'s "Ten Worst Britons" list of 2005, a group that includes Jack the Ripper.) As long as threats to the Church were viewed synonymously as threats to the State, the Crown would legally enforce these spiritual rulings.

Despite the fierce opposition by the two most powerful organizations of the time, Wycliffite Bibles were so widely distributed and cherished that around two hundred fifty still exist today, even after a lengthy purge to burn them out of England. In comparison, only about sixty manuscript copies of Chaucer's masterpiece of English literature from the same century, *The Canterbury Tales*, survive.

Of course, it wasn't just the Bibles that the Catholic Church was interested in burning. As keepers of Christ's flame, they felt a duty—nay, an imperative responsibility—to use that flame to melt people who did not agree with them. The idea that "simple folk" should be able to read the Scripture in their own tongue was just such a point of disagreement.

MODERN READERS have a hard time understanding the danger in allowing everyday folk to read books, especially the foundational texts of their own religion. But, for the strictly orthodox, widespread scriptural literacy was inextricably intertwined with chaos. The Wycliffite movement was gaining steam at the same time as the Peasants' Revolt of 1381, a shocking and bloody uprising in which rebels marched on the English capital and slew a number of royal officials. Wycliffe and his followers were not the cause of the uprising; London had the Black Death and taxes to thank for that. Religious fights over books, however, were really fights over foundational structures of power—and this right at the time when rebels were storming the Tower of London. Amid these struggles, many defenders of orthodoxy came to believe that heresy was inevitably intertwined with sedition.

Some of these reformers would be gathered up and required to cast their heretical English Bibles into great bonfires. Others were given closer ringside seats: in 1496, five resurgent followers of Wycliffe's Lollards were burned at the stake in London with their beloved English New Testaments tied around their necks. At the time, a heretic hadn't been burned in England for two hundred years—the last was a deacon who'd converted to Judaism—but the threat of an English Bible was enough to dust off the pyres and reopen that chapter of human barbarism. Let's be clear: there is no such thing as a "good death" when it comes to immolation. John Foxe, the most famous of martyrologists, recorded the passing of one "heretic" who took forty-five minutes to expire in the flames. Seeing that his "left arm was on fire and burned, he rubbed it with his right hand, and it fell from his body, and he continued in prayer to the end without moving."

Despite facing a horrendous death, courageous men and women refused to budge on issues that seem trivial to many modern observers. For example, does the wafer in Communion literally turn into the flesh of a man who lived fifteen hundred years ago when it touches your tongue? If you marked ❏No, ❏Probably Not, or ❏ Undecided, congratulations, you would have been burned at the stake. While tied to just such a stake, John Badby used his last words to affirm that "he knew well it was hallowed bread, and not God's body."

What separated the times of William Tyndale most markedly from those of John Wycliffe was not the burning of heretics—there was plenty more of that to come—but, rather, the reach that the printing press provided. While presses were being tightly watched in England, printed contraband was making its way from mainland Europe into the realm of King Henry VIII by means of robust smuggling networks.

In the spring of 1524, Tyndale realized that England was about the worst place he could be if he wanted to create an English Bible. He made his way to Europe and, using illegal shipping routes forged by capitalist innovators and Lollard heretics alike, spent the next decade sending a stream of inflammatory words right back into the heart of England.

TYNDALE'S FIRST target on this crusade was his most dangerous. Nothing else that he printed would cause more suffering, result in more torture, or send more Englishmen to the flames than the life and words of Jesus of Nazareth.

Because he was soon to be branded an international fugitive, it's unclear where Tyndale was located while he translated the first edition of his English New Testament. We do know where he printed it, though. Well, at least where he *attempted* to print it. In 1525, Tyndale negotiated with a printer in Cologne for three thousand magnificent copies of his newly translated New Testament. Using a Greek edition (the original language of the New Testament writings) created by the internationally renowned scholar Erasmus nine years before, along with Martin Luther's German translation and the Latin Vulgate, Tyndale's version was to be the most complete, the most accurate, and the most beautiful rendering of God's word into the English language ever smuggled into the country in bales of cloth.

Technically it was illegal to print the Bible in English, even in Europe. But Tyndale managed to secure a verbal contract for a hefty run of finely crafted New Testaments. Printed on decadently large quarto pages with comments and glosses to guide the reader, a detailed prologue, and woodcut illustrations, this would be the kind of Bible that any God-fearing Christian would be proud to display at home. From an "aesthetic" perspective, that's great. From an "evidence used to burn you alive" perspective, it's a bit impractical.

Tyndale didn't make it very far with this first printing before a friend and correspondent of Bishop Tunstall's showed up at the print shop. Johannes Cochlaeus was a hard-line Catholic and known

English Bible hunter. With exceptionally terrible timing, Cochlaeus traveled to Cologne in 1525 looking for someone to print the works of a conservative monk and ended up on the doorstep of the man who was at that very moment printing Tyndale's English New Testament.

But hey, just because a printer is playing both sides of the fence doesn't necessarily mean his polemically opposed clients will come in contact with each other. As it turns out, however, sixteenth-century printing house employees were the stereotypical construction workers of their day. They worked long and dangerous hours, and treated that stress with lots and lots of alcohol. Inebriation has been known to produce the following effects in human beings: speaking in a raised voice, laughing too hard at things that aren't funny, being sleepy and uncoordinated, and telling people in taverns that no matter how much kings or cardinals rage, "all England would in short time be Lutheran."

Cochlaeus was most interested in that last one. So he invited the printers back to his place for another round of drinks, on him. Hurrah! In gratitude, they let slip that they were working on an English New Testament, that there would be three thousand copies, that those copies would be smuggled into England, and that they were about eighty pages into the project. Cochlaeus was "inwardly astonished and horrified," and soon began making arrangements to seize the contraband Bibles.

Luckily, Tyndale was tipped off, so he was able to scoop up his Bible (most of the Book of Matthew, anyway) and flee Cologne. Eventually he arrived in Worms, a German city originally named after the Norse hero Siegfried, who was said to have slain a lindworm (a wingless dragon) near its borders. More recently for Tyndale, Worms was the site of Martin Luther's trial in front of an imperial diet in 1521. At its conclusion, Luther was officially discredited, declared a "devil in the habit of a monk," and watched as his books were officially banned. Three years later, however, the tide turned in Worms and it became a safe haven for Lutheran printing.

This time around, Tyndale played the part of fugitive seditionist better. In his first printing attempt, he conceived of a flashy quarto, a somewhat large format for a book, made when each sheet taken through the press is folded twice before binding, creating four printed leaves per sheet. In this format, the book generally ends up just a bit bigger and squarer than your average hardcover release today. Now Tyndale opted for a more subdued octavo, a book format that required folding the printed sheets once more before binding, thus producing eight leaves per printed sheet—and a smaller book. His octavo could fit in a pocket, making it easy for anyone to carry, read, and study it—also to smuggle, conceal, and sell it on the black market.

There were no prologues to these smaller Bibles, no comments or glosses, few woodcuts, and no colophon. The first complete New Testament ever to be printed in the English language was created in Germany by a fugitive linguist and a rogue printer, who took great pains to cover their literary tracks. They were so successful, in fact, that the only complete copy of Tyndale's 1526 New Testament wasn't identified until 1996. In Tyndale's time, the octavo Bibles could be purchased for six days' wages of an unskilled worker. In 1994 the British Library purchased an incomplete copy for "a little more than one million pounds." The Library has noted with chagrin that when a famous collector "bought it . . . in the 1700s, it had cost him just 20 guineas."

Tyndale spent more than a decade on the run. He published two editions of the New Testament and a translation of the Pentateuch (the first five books of the Old Testament), and he wrote, printed, and distributed numerous books, Testaments, introductions, and letters in defense of his person and beliefs. He moved around the Low Countries and was cautious about meeting anyone in person.

He would not sit for portraits because they might be used to identify his face. As a by-product, almost five hundred years later, we don't know what Tyndale looked like. There is a portrait hanging in a dining hall at Oxford University with an inscription that reads, "This picture represents, as far as art could, William Tyndale." Translation: "Here's a painting of a white European guy. Please pretend this is William Tyndale."

Was Tyndale just being paranoid? Well, judge for yourself: William Tyndale is "a hellhound in the kennel of the devil." He discharges "a filthy foam of blasphemies out of his brutish, beastly mouth." His English translation should be called "the testament of his master, Antichrist." Tyndale "sheweth himself so puffed up with the poison of pride, malice, and evil, that it is more than a marvel that the skin can hold together."

These elegant pronouncements fell from the pen of the aforementioned statesmen Thomas More, who would prove to be Tyndale's greatest enemy and the single most dangerous man in England next to beheading-obsessed Henry VIII. Whereas other Catholic zealots were content to humble the heretics, forcing them to recant on bended knee, More brought to bear all the powers of Church and State to hunt down these "demonic agents" and send them to the stake.

Three years after Tyndale's New Testament first spread throughout England, Thomas More became the country's lord chancellor. A career politician, More was already famous for his brilliant and witty international best seller, *Utopia*, a satirical exploration of the perfect political society. Of course, you had to read Latin to gain literary entrance into that society; More prohibited *Utopia* from publication in English "lest it might fall into the hands of the simple and unlearned." Even today, in some ways, access to this publication is limited to the uppermost classes of society. Early printings appear on the market these days for around $50,000 to

$100,000, while the 1516 first edition last sold at auction, in 2005, for €210,000. To More's posthumous dismay (we're guessing), the first edition in English of his best-known work also commands prices of $50,000 to $100,000, even though it's much later than the early Latin printings. It turns out that collectors actually like to *read* the text of the books they buy, and educated bibliophiles of the twenty-first century don't often read Latin anymore.

More was one of the most strident believers that heresy was a threat to an orderly and civilized society. He felt that conflicts such as the Peasants' Revolt demonstrated how, when they were given access to too much information, the common people inevitably became a murderous mob. He could even point to evidence more recent than the 1381 English revolt: in 1524–25 a massive uprising occurred across Germany. Some of the leaders of this Peasants' War cited Luther's rebellion against the Church as a key influence. With More's new promotion to primary adviser to Henry VIII, he enjoyed immense political and judicial power. In particular, it was the duty of the chancellor to protect the state from treason, sedition, and rebellion. Bible heretics were a threat not just to the Church, but also to the State itself. More held the keys to the racks and the pyres of England, and in the name of civilization, he developed an iron stomach for burning men alive.

In the streets of London there were plenty of heretical activities that could be branded treasonous. According to one scholar, "Destroying images, posting bills, singing seditious ballads, spreading forbidden books, hiding those on the run from the authorities, taunting priests, meeting in secret conventicles, planning daring escapes, preaching in spite of persecution, the first followers of the new doctrines marked themselves as rebels as well as heretics." Lacking only the asthmatic mask and red light saber, More became singularly obsessed with hunting down that rebel alliance.

In the twenty-five years that he served in the English Parliament, More had learned well the power of the printed word. Around 1521 he edited (and to some extent authored) Henry VIII's attack on Martin Luther, entitled *Assertio septem sacramentorum* (*Defense of the Seven Sacraments*), which won Henry the title of *Fidei defensor* ("Defender of the Faith") from the pope. Martin Luther didn't care at all for that *assertio*, so he published one back, stating that Henry was "a pig, dolt and liar who deserved, among other things, to be covered in excrement."

King Henry knighted Thomas More that same year and asked him to respond to Luther on his behalf. Presumably More would be free to print the types of things that were below the station of the king of England. Here is a highlight reel from *Responsio ad Lutherum*, published in 1523:

"Martin Luther is an ape, an arse, a drunkard, a piece of scurf [dandruff], a pestilential buffoon, and a dishonest liar . . . Someone should shit into Luther's mouth, he farts anathema, it will be right to piss into his mouth, he is filled with shit, dung, filth, and excrement; look, my own fingers are covered with shit when I try to clean his filthy mouth."

Wow. One eighteenth-century commentator described More's tirade as "the greatest heap of nasty language that was perhaps ever put together." That man had never seen a YouTube comment section, but point taken. To make matters worse, More was a saint—not at the time, obviously, but he would be canonized in 1935 and later, in 2000, declared the patron saint of statesmen and politicians by the Vatican. This is the same man whom scholars stop to examine for his "obsessive anality." Congratulations, Saint Thomas, there are very few saints from whom Latin students can learn such a variety of filthy Renaissance curse words.

Between 1526 and 1535, Thomas More and William Tyndale declared war on each other. Their battles didn't take place in fields and valleys, but on the pages of the printing press. It was a power struggle embodying the conflict of the whole of England. In order to convince the British people that he was right, More even broke form and began publishing in the common tongue. And when he jumped into the English printing pool, he cannonballed, producing a staggering half-million-word corpus of vitriol aimed at Tyndale and other heretics. Between these two men, three-quarters of a million words were written, producing one of the greatest theological debates ever printed in the English language. More and Tyndale bickered over the usual topics, such as the role of the pope, and the king's love life, but their main fight centered on the printing and distribution of Tyndale's English-language Bible, the humble octavo that had come to divide the powers of heaven and earth.

MORE CLAIMED that he would support a carefully screened English translation of the Bible, but that Tyndale's translation contained as many errors as the sea had water. It is worth noting that the committee of scholars tasked by King James I to create the authorized version of the Bible in 1611 found that Tyndale's translation was almost entirely accurate and faithful to the original texts. In reality, More's beef with Tyndale's New Testament came down to three words. Out of 138,000 words describing the life and works of Jesus of Nazareth—three were objectionable. Yet, to More, anything above zero was enough to make men burn.

Congregation. Elder. Love.

Congregation, a seemingly harmless word, in fact undermined an entire power structure. This word appears only three times in the New Testament, but its placement had profound implications for the pope. In a particular verse in the Book of Matthew, Jesus appears to be making his disciple, Peter, the steward of his "Church." For hundreds of years, popes claimed to have inherited this charge, thereby serving as the rightful heirs to the leadership of all Christianity.

By changing the word *Church* to the more abstract (but equally accurate) *congregation*, Tyndale dissolved that papal authority. In a single word, the pope was created, and in a single word the pope could be erased. This was not a trivial translation choice, either, but a serious power play. An entire sect of Christianity, Congregationalists, subsequently sprang up around this word.

Tyndale also chose to change the word *priest* to *senior* or *elder.* This created obvious threats to the current authorities. Should the Church be run by priests and bishops, or by senior members of the congregation (presbyters)? The Presbyterians arose from this word's translation.

There is a slight linguistic difference between the words *love* and *charity*, but theologically, they are light-years apart. Tyndale argued that New Testament "love" was unattached to good works, or gifts, or charitable donations that would flow to the clergy and the Church. Because of this debate, the lord chancellor of England came to view the word *love* as synonymous with "heresy."

Goodnight, my husband More's wife could have said. *I love you.*

What did you say, Alice?

Oh, right. I meant, good night, husband. I feel charitable toward you in a way not independent of good works or the rightful claim of papal authority over the believers of Christ.

Simple words are not so simple. How we interpret a word depends on who says it, and why. The reformers whom More called "seditious" were using these words as weapons against those in power. They were proverbial torches thrown through the window of the Church, threatening to destroy it from the inside.

IN 1529, the same year that More was elevated to the lord chancellorship, Tyndale made his last move, to Antwerp (in modern Belgium), a haven for Protestant writers. Along with Paris and Venice, Antwerp was an international mecca for printing. It was a thriving trade city, with large amounts of capital and solid distribution networks. From the ports in Antwerp, books were being smuggled across the North Sea into England. Tyndale could blend in, write to his heart's content, and, as long as he wasn't too brazen about his illegal activities, relax a bit. After four years on the lam, his new lifestyle was becoming second nature. Sir Thomas Elyot, an ambassador from England, launched an eight-month investigation attempting to arrest and deport Tyndale, but he came up empty-handed and deeply in debt. "As far as I can perceive," he wrote to the Duke of Norfolk, "hearing of the king's diligence in the apprehension of him, [Tyndale] withdraweth into such places where he thinketh to be farthest out of danger." The frustrated ambassador soon resigned his position and returned to England.

Henry VIII was not happy with Tyndale, even despite Tyndale's public support of him as the highest authority in England, surpassing the pope. This was important because, at the time, Henry and Pope Clement VII were in serious disagreement over the king's marital affairs.

More than twenty years prior, Henry had married Catherine of Aragon, a Spanish princess. Catherine was an unusually competent Euruler. She previously served as the first female ambassador in European history, played a major command role in the Battle of Flodden (the largest between England and Scotland), and championed women's education and care for the poor. Yet, no matter how adept she was in the field or at the negotiating tables of Europe, "no other success can compensate for failure in the home." And by that standard, Catherine was a goddamned disaster. Forget sensitive compromises with Spain or giving food to the needy; if you can't make male babies that live longer than a month and a half, your value as a woman is forfeit. So, the king went looking elsewhere for a male heir, and he found that elsewhere in the fancy bits of the ambitious twenty-four-year-old Anne Boleyn.

The wooing, engagement, and eventual royal marriage to Anne Boleyn was known as the king's "Great Matter." The Catholic Church refused to recognize Henry's marriage annulment. Making up euphemisms such as "Great Matter" or "fancy bits" didn't seem to make much difference, either. Resistance from the Church made it exceedingly difficult for someone to drop his old wife and run off with a younger, prettier secretary . . . er, lady-in-waiting. Though it was framed in admittedly more diplomatic terms, it was over this issue that Henry VIII and Pope Clement VII locked themselves in a struggle that would transmute the religious landscape of England.

Already fiercely antipapal, Tyndale was only too happy to declare the word of the king above that of the pope. But he was also anti-leaving-your-wife-for-a-younger-woman, which put him in a bit of a moral quandary.

Anne Boleyn knew of William Tyndale. In some ways, you could say that she was a fan. When you're engaged to a king like Henry, you don't give two Thomas More shits which books are banned by the bishops of England. You read whatever you want, whenever you want. And Anne apparently had connections in the smuggling market. Today, the British Library owns her personal copy of Tyndale's 1534 New Testament, which was not only perused at her leisure but also reportedly "kept . . . open on her desk in her chamber, [and] which her servants were all encouraged to read."

Anne consumed Tyndale's works, including *The Obedience of a Christian Man*, which, among other things, told the Catholic Church to keep its grubby mitts out of sovereign state business. Not surprisingly, Anne was particularly partial to this tract, and some accounts say she presented Henry with her favorite passages to read. At one point, Henry even reached out to Tyndale, asking him to give in on a few theological points and return to England in the king's full favor. Tyndale agreed to do so—in fact, he agreed never to write another polemic again—if the king would only allow an English translation of the Bible. But that, apparently, was too much for Henry to promise—for the time being, anyway.

William Tyndale was a revolutionary and a highly intelligent man, but he was also entirely uncompromising in his views. While this moral fiber makes for honorable Reformation heroes, when it comes to politics it makes for lumbering clodpates. Rather than embracing his immensely powerful new ally, Tyndale could find no justification in the Bible for divorce, and in 1530 he published *The Practice of Prelates*, wherein he publicly denounced the union between Henry and Anne. In a meeting with the king's agent, Tyndale expressed surprise at Henry's reaction: "I am informed that the king's grace taketh great displeasure with me . . . specifically for the book name *The Practice of Prelates*: whereof I have no little marvel, considering that in it I did but warn his grace of the subtle demeanor of the clergy." A little political sense would have gone a long way for Tyndale. Of course, "long way" probably should be read as "six years." After repeating Catherine's mistake of failing to produce baby boys, Anne herself was cast aside for a newer model. Her marriage to Henry was declared the result of witchcraft (sortilege), and after charges of adultery and incest floated around the English courts, Henry officially separated from Anne by separating her head from her body.

THOUGH BITTER enemies, William Tyndale and Thomas More did agree on one point: Henry's divorce from Catherine of Aragon was a crime against God. This was a dangerous position for anyone to take, but especially the king's closest adviser. The danger was compounded exponentially when Henry finally broke from the Catholic Church and declared himself the "Supreme

Governor of the Church of England." This step was just too far for More. He resigned as lord chancellor, and like Tyndale, his life was now in danger for his religious beliefs.

More's resignation didn't necessarily stop his heretic crusade. This was a man who had built up a shadow army of informers that stalked the streets of London at night searching for that most pernicious of crimes against humanity: Bible study groups. Anyone who's had to sit through a particularly bromidic Sunday school instructor might be inclined to agree, but in the early 1530s, those tedious lessons would have been the very height of political intrigue. Groups with heretical sympathies would meet at night to read aloud from English translations of the Bible, a criminal organization that one scholar calls the "Forbidden Book of the Month Club." On the other side of the candlelight-flickering walls lurked More's network—effective, brutal, quietly stealing names and rumors and unshielded comments in the dark.

Charges of heresy were not easy to beat. The accusers were often anonymous, and the definitions of "heresy" elusive and subjective. There *were* laws that protected heretics, which dated back to Henry IV in 1401. For example, a suspect could be held for only ten days before being delivered to an ecclesiastical authority. A bishop couldn't hold a suspect for more than three months before trial. If the accused was acquitted, he could not be arrested on the same charges again. Six weeks after becoming lord chancellor, however, More took those laws and figuratively punted them into the English Channel. Facing rumors of torture and indefinite confinement, some people recanted and surrendered themselves to whatever punishments More could bring to bear.

Thomas More didn't burn *everyone* he detained. As a heretic, you had one mulligan: recant. But if you cashed in your single recantation and were caught a second time—off to the flames with you, friend. Not that your heresy mulligan was a walk in the park. Well, in some cases it was a walk in the park, but that park was located at Thomas More's house.

More's home stood in Chelsea, located in central London. It was an impressive estate, containing a library, private chapel, ample guest accommodations, a menagerie, even a dock for More's personal barge. It had lavish gardens, with "apple trees, roses, and a collection of herbs." Yet within this garden also stood the "Tree of Truth." Despite its name, this tree wasn't in the habit of dropping apples of wisdom. It was more into the extraction side of the business.

Rumors began circulating that More was illegally detaining suspected heretics and torturing them on his property, presumably in an attempt to make them confess. He wholeheartedly denied the allegations. Sure, the gatehouse on his estate was equipped with stocks and chains and fetters, but c'mon, those were collectors' items. Everyone knows an Iron Maiden is great Feng Shui, and a person's choice of décor isn't a crime, is it?

More did admit to arresting and detaining a merchant named Segar Nicholson, for selling contraband books. Nicholson was among the first to accuse More of torturing him at his home in Chelsea. He claimed that the lord chancellor caused him to be bound to the Tree of Truth, whipped, and then subjected to ever-tightening ropes around his skull until he lost consciousness. More dismissed the accusation, saying of Nicholson, "never had either bodily harm done him or foul word spoken him while he was in mine house." A known seller of antichrist books was in your custody for a week and you didn't torture him even a tiny bit? Not a single screw to a thumb, or a morning's stretch on the rack? Sounds like someone else could use a visit to the Tree of Truth.

After he resigned his post as lord chancellor, Thomas More's days were numbered. He still sought out heretics wherever he could find them, but his political support was withering fast. Henry was angry with him, Anne Boleyn held nothing but contempt for him, and his allegiance to the pope in Rome was becoming more and more problematic. It was only a matter of time before charges of treason found him. This happened in April 1534, when he was brought before a commission to take the "Oath of Supremacy," which recognized Henry VIII as the supreme head of the Church. To refuse was treason, and on that day, Sir Thomas More became a traitor to England.

More was charged with depriving the king of one of his titles. He was sentenced "to be hanged, cut down while still alive, castrated, his entrails cut out and burnt before his eyes, and then beheaded." Henry, the ultimate frenemy, commuted the sentence to just decapitation. In July of the following year, More was executed. His head was boiled and stuck atop a pole on London Bridge. Had his daughter not bribed the constable of the watch, this last vestige of More would have been unceremoniously dumped into the river below.

AS THOMAS More's falling out with the English monarchy became increasingly lethal, Tyndale was heavily engaged in his most important contribution: the second edition of his English New Testament. Published seven months after More was declared a traitor, Tyndale's 1534 New Testament would become the gold standard of biblical translations. This more refined and accurate version spread like wildfire, revolutionizing not only religious study, but the English language itself.

The Bible is the most popular book printed in English. Tyndale's New Testament was the great trailblazer, cutting paths through the unruly wilderness of the English language. His authoritative biographer, David Daniell, goes as far as to say that Tyndale's work was a "liberation of language itself," and without him, there could have been no Shakespeare: "I cannot express too strongly the revolutionary effect of this release of rhetoric into English . . . Something happened, before Shakespeare, to switch the power into English. What happened, I am sure, was the constant household reading of the Bible in English." The translations that those households were reading owed the majority of their text to Tyndale. The Geneva Bible, the translation Shakespeare used (as did Milton, as did the Pilgrims), relied on the path blazed by Tyndale. So many of the phrases we know by heart ("let there be light," "the spirit is willing," "there were shepherds abiding in the field") came from his mind. He was an extraordinary translator above all else, weaving a text of such beauty that, hundreds of years later, we still frequently stop to admire it.

Unfortunately for Tyndale, revolutionizing the English language is not a guarantor of personal safety. A couple of months before his nemesis was beheaded in London, Tyndale received a visitor to his home in Antwerp. The young man was in his twenties, came from an affluent family in Dorset, England, and had graduated from Oxford with a bachelor's degree in civil law. He was also a gambler, a notorious thief, and a conman. And like most conmen, Henry Phillips did not break into Tyndale's home but, rather, was invited in by the victim himself.

Tyndale had been on the run for roughly a decade by now, but the great wheels of change were turning all over Europe. Through the influence of Anne Boleyn and the new lord chancellor, Thomas Cromwell, England was becoming friendlier to Protestants every day. But the Low Countries, where Tyndale was hiding, were marching toward an ever-more-frightening state of conservative Catholicism. In 1533, for example, an anonymous letter circulated among government authorities that complained about the lack of an organized Inquisition "in the Spanish manner" to hunt heretics. Raids to arrest heretics followed soon after in the very city where Tyndale resided. His own name ended up on the list, and he had to work harder than ever to stay one step ahead of the flames.

Into these changing tides stepped Henry Phillips, who met Tyndale at a dinner for Antwerp merchants in 1535. Tyndale instantly took a liking to the well-spoken young man and invited him back to his lodgings to continue their conversation. As the two men talked into the night, Tyndale did not realize that his new BFF harbored some seriously dark secrets. Among them was the fact that he had been hired to abduct Tyndale and turn him over to Catholic authorities.

In Antwerp stood the English House, a kind of embassy for English merchants. It was here that Tyndale had sought refuge. In line with its sovereign status, the city of Antwerp couldn't just break into the English House and start dragging out international suspects. Tyndale had to be lured outside, and it was Phillips who became the bait.

We wonder if one of the most important translators of the life of Jesus looked back and caught the similarities: a cadre of officers waiting in concealment, a dinner appointment, a nod of the head in place of a kiss, a substantial monetary reward. When Tyndale realized what was happening, he simply gave himself up without a word or a struggle. After a decade on the run, it was over.

Ever since Tyndale's betrayal, there has been considerable debate over who orchestrated his arrest. As far as we know, Phillips never revealed the name of the person or persons who sent him to hunt down the rogue translator. Investigations at the time placed blame on various bishops in England, but the accusations are full of dead ends and contradictions. History may never know the identity of the man who doomed Tyndale, but it does know certain things *about* the man. For one, to cover his expenses and necessary bribes, Henry Phillips would have needed access to a considerable amount of money. His shadowy benefactor would also have been familiar with the Low Countries and their laws and procedures. In order to dig up the information about Tyndale's whereabouts that led to Phillips's "friendly" approach, a robust system of informants would have been necessary. Above all, despite the tides that had turned so dramatically in England, this person must have felt nothing but a continuing and burning hatred for Tyndale and his work.

Not many people in England fit the bill. Some might note that, up until the very last stages of his imprisonment, Thomas More was free to write and coordinate with friends on the outside. If Tyndale's greatest nemesis wasn't behind his final abduction, it was another person, his name lost to history, who closely matches Thomas More's profile.

The betrayer Henry Phillips pops up a few times around Europe before dropping off the historical record entirely. He surfaces in Rome a year after his betrayal of Tyndale, asking for money and claiming to be kin of the now-executed Thomas More. His later years were reportedly steeped in paranoia and poverty, and he bounced around such promising careers as mercenary, highwayman, and beggar in the streets of Vienna. The martyrologist John Foxe recorded that Phillips was "consumed at last with lice."

During the year and a half that Tyndale was imprisoned at Castle Vilvoorde (in modern-day Belgium), numerous attempts were made by friends and government officials to secure his release. Even Henry VIII asked for leniency, and the powerful Thomas Cromwell pressed tirelessly. While those petitions caused delays, in the end they fell on deaf ears. Heresy was an international offense, and Henry's power was limited outside England. Also, Emperor Charles V, who ruled over the Low Countries, was Catherine of Aragon's nephew. Catherine had died just three years before, dishonored and forsaken by the King of England. Charles wasn't exactly in the mood to extend Henry any personal favors.

While Tyndale was languishing in prison, another English translation of the Bible was being prepared, by Tyndale's friend and colleague Miles Coverdale. In his edition, Coverdale made sure to play nice with Henry VIII, noting in the dedication that "power given of God unto kings is in earth above all other powers." In 1537, less than twelve months after Tyndale died, a new edition of Coverdale's translation became the first English Bible officially licensed by the king. The translation was, of course, mostly Tyndale's. However, the name of a heretic couldn't be ascribed to a royal work such as this, so it was called the "Thomas Matthew Bible," after two New Testament apostles. To pass the king's inspection, Tyndale's contribution had to be buried.

Like his greatest enemy, Thomas More, William Tyndale was condemned to die for his religious convictions. This reportedly happened near Vilvoorde, on October 6, 1536. Unlike More, whose sentence was reduced to beheading, Tyndale did not die easily. Because it couldn't be proven that he'd been given his one heretic mulligan, the Belgian courts decided to extend to Tyndale the mercy of strangulation. While choking someone with the intent to kill normally isn't considered a mercy, when the alternative is live immolation, it's a godsend. Unfortunately for Tyndale, God failed to deliver on that one. If contemporary accounts are accurate, the strangulation was botched, and Tyndale was burned while yet alive and speaking. That would mean the author of one of the most important works ever printed in the English language was forced to endure a horrific death not once, but twice.

On the back of the printing press, Tyndale fundamentally altered the religiopolitical landscape of England. Small-scale English translations of the Bible had come and gone for more than a millennium, but Tyndale produced his greatest creation by virtue of Gutenberg producing his. Seventy-five years later, when the monumental King James Version of the Bible was published, Tyndale's translation would account for 83 percent of its New Testament and 76 percent of its Old.

It is a cruel irony that the crime for which Tyndale was hunted, abducted, and brutally executed became one of the greatest English accomplishments within a few decades after his death. Perhaps, when you get yourself mixed up in one of the largest power struggles in English history, things like that are bound to happen. Today, when people blame a "scapegoat," or encourage friends to "fight the good fight," or compliment "the salt of the earth," we can thank William Tyndale. Ultimately, though, when you're hunted by religious fanatics, bloodthirsty emperors and kings, and an eloquent nemesis with a scatological fetish, even the printing press can't prevent you from becoming the "forgotten ghost of the English language."

CHAPTER FOUR

4

MAKING THE ROUND WORLD FLAT

IN THE SIXTEENTH CENTURY, EUROPE found itself in the midst of a trade route arms race. Everyone was assembling maps and expeditions to stake their claim on the economic arteries of the world. The voyages of Christopher Columbus helped Spain call dibs on the Americas. Sailing around Africa to reach India was Portugal's gambit. Italian city-states such as the Republic of Venice dominated many other routes to the East. England, however, had nothing. It was the scrawny kid picked last for imperialist dodgeball.

For decades, British explorers made several attempts to own the Arctic. More than anything else, they wanted to break through the frigid lands and waters of northern Russia to establish profitable trade routes with the Far East. Anyone who is slightly familiar with the Land of Rus or its ability to churn out an ungodly amount of snow and ice would peg this for the fool's errand that it was, but English politicians and explorers had an ace up their sleeve: a map printed in 1569 by the most famous cartographer alive, Gerard Mercator.

MERCATOR'S MAP OF THE NORTH POLE, *SEPTENTRIONALIUM TERRARUM DESCRIPTIO*, FIRST PUBLISHED IN 1569. THIS IS A 1611 IMPRESSION OF A 1595 UPDATE, *courtesy of Leen Helmink Antique Maps.*

This is a top-down view of the Arctic Circle as it was understood by Mercator and spread by printing presses in the late sixteenth century. The first thing that should strike a modern observer is its sheer beauty. If not, please get your AI software updated, because you are not human. The second observation is the presence of four giant islands, each roughly the size of three Texases, sitting in the middle of the Arctic Ocean. Unlike its southern cousin, the

Arctic is essentially void of land mass. Did people in the sixteenth century know that? No. How could they? Humanity didn't fully grasp the emptiness of the Arctic until we could take pictures of it from orbit.

The great mapmaker Mercator didn't have access to low-earth-orbit satellites, but he did have access to something nearly as effective: hearsay. In the library of one of Mercator's friends sat the manuscript of Jacobus Cnoyen, a world traveler who'd lived two centuries before. In 1364, Cnoyen visited the court of the king of Norway and met with a very interesting priest. The priest claimed to be descended from a group of four thousand settlers sent to the Arctic by none other than King Arthur. The account of this settlement was contained in a document known to Cnoyen as the *Gestae Arthuri*, which is now lost.

The priest also informed Cnoyen that a monk from Oxford had recently visited the islands and met with the Norwegian king personally. What did they discuss? That the monk, armed with an astrolabe, had wandered around the Arctic mapping everything in front him. He'd "put into writing all the wonders of those islands" and now gave King Edward III the book, called *Inventio Fortunata* ("Blessed Discovery"). This book is also, you guessed it, lost.

It was from this wandering monk that a priest learned about the topography of the Arctic Circle. The priest then passed that information to globetrotter Jacobus Cnoyen in 1364, who wrote it in his travelogue, which landed in the library of Mercator's friend in Antwerp (and was lost sometime in the 1580s). In 1577, Mercator sent a transcription of that travelogue along with a summary of its importance to his friend John Dee, who was a personal adviser to Queen Elizabeth I.

This was all fantastic news for Britain for two very important reasons:

1. Finders keepers. Whoever gets to a new place automatically owns that place by virtue of planting his flag first. This is why America, um, legally owns the moon. If Arthur settled the Arctic Circle in the sixth century, then the Arctic Circle naturally belongs to England. Including . . .
2. The Northwest and Northeast Passages. Look at that map again. Even with the naked eye, you can see that it's entirely possible to sail a ship from England to China just by skirting those four big islands. You might even slow down when passing one of them to wave at its "pygmies whose length in all is 4 feet." Or at the imaginary inhabitants of another island, described as "the best and most salubrious of the whole of Septentrion"—which sounds like made-up Harry Potter words but actually meant "the healthiest folks of the northern regions."

Mercator's famous name was wielded by those who supported an English claim to imaginary trade routes in the New World. Carrying his 1569 world map, English explorers of the late sixteenth century, such as Martin Frobisher, smashed themselves against the ice-locked lands of northern Canada for years (not without casualties) before finally giving up and returning home. (At least they didn't return empty-handed. Frobisher hauled literally tons of Canadian gold back to England. Canadian gold in this case being iron pyrite—ye olde fool's gold.)

LIKE AN Englishman into a shelf of Arctic ice, Europe crashed into its first information age in the 1500s. This deluge of raw data was made possible by three main factors: the Age of Discovery, humanism, and the rise of printing. A flood of information like this comes with unique challenges. A flood is not picky.

It pulls in trash and mixes it with treasure. Sometimes, in that massive vortex, it's hard to distinguish between the two. Gerard Mercator made a valiant effort, but even the great cartographer wasn't immune to misinformation. We shouldn't judge him too harshly, though. He lived in a time when the face of the planet was changing every day.

The Age of Discovery was the time of famous explorers such as Christopher Columbus, Amerigo Vespucci, Vasco da Gama, Magellan, Hernán Cortés, and Sir Francis Drake. George Beste, an associate of Frobisher's, said that within "fourscore years, there hath been more new countries and regions discovered than in five thousand years before." As these adventurers returned home (often having raped and pillaged their way through entire continents), their accounts were recorded in vibrant detail (sans the rape and pillage) and fed, with varying degrees of accuracy and subterfuge, into the ravenous printing presses of Europe.

Intrepid humanists began (re)discovering ancient manuscripts, and brought them out of obscurity with the help of those same printers. Ann Blair notes that the driving force behind it all was the Renaissance "info-lust that sought to gather and manage as much information as possible." In a way that had never been possible in human history (and wouldn't be again until the Internet), Europeans suddenly had access to an explosive amount of new information.

However, not everyone agreed that this flood of knowledge was beneficial. Some writers had been complaining for centuries that we already had "too many" books (a concept we fundamentally don't understand, but that was nevertheless quite widespread). In the first century CE Roman philosopher Seneca declared that "the abundance of books is distraction." Erasmus, the sixteenth-century scholar whose fame was spread far and wide by printing presses, infamously said, "Is there anywhere on earth exempt from these swarms of new books?"

The success with which Europe handled its first information age is decidedly mixed. On the one hand, it forced writers to develop more advanced methods of containing and organizing information. On the other, even writers with the best of intentions were sometimes unable to distinguish between trustworthy sources and (in retrospect) preposterous ones. More than anyone else, the perceptive and talented Gerard Mercator embodied this absurd challenge.

Mercator's story really begins more than a thousand years before his birth, with a Greco-Egyptian geographer named Claudius Ptolemy. Ptolemy was to mapmaking what Isaac Newton would be to physics: the foundation. In the second century CE, Ptolemy tried to delineate a scientific geography based on observation. He adapted lines of longitude and latitude that stretched out over the world, dividing and organizing the globe so that coordinates could be taken and distances properly measured. Two dozen hand-drawn maps contained the entire planet, neatly described in Ptolemy's groundbreaking work, the *Geographia*.

So important was the *Geographia* that fourteen centuries later, Mercator would build on it to spark a golden age of cartography. *Fourteen centuries later.* This is like ARPANET being created in 1977, but the Internet not existing until the year 3377. There is a fairly good reason for its taking fourteen hundred years to advance from Ptolemaic maps to Mercatorial maps. After the fall of the Roman Empire, Ptolemy's work was lost and essentially forgotten in Europe. Luckily for civilization, Arab scientists were experiencing their own golden age, during the European "dark ages," and were using and preserving manuscripts of Ptolemy's work. If it weren't for them, we may have lost Ptolemy entirely.

The *Geographia* was translated into Latin from a Greek manuscript as early as 1406, and Byzantine scholars began sharing it more widely with their Italian counterparts after the fall of Constantinople in 1453. However, it wasn't until the invention of the printing press that Ptolemy's work was introduced to Europe on a mass scale. With the Renaissance's lust for information, the ancient geographer's maps were re-created and published all over the Western world, ushering in a reawakening of cartographical methodology. The *Geographia* was first printed in Rome in 1469 by Sweynheim and Pannartz, two German printers who introduced the art to Italy. Today this edition of *Geographia* is incredibly rare: in the past seventy-five years, only one copy has ever appeared at auction. If you were to find a copy in your attic—you won't, but let's just say—you would probably be able to sell it for close to half a million dollars.

The Age of Discovery went hand in hand with the Ptolemaic revival. Columbus carried a copy of the *Geographia* on his first transatlantic voyage. The fact that he was carrying it with him in 1492 should raise a red flag, as his voyage is not necessarily a shining example of geographical understanding. On the reconstructions of Ptolemy's maps, Africa resembled an amorphous blob rising out of a mythical southern continent, and Asia was close enough to Spain for Columbus to (as Sarah Palin might say) see it from his house.

Ptolemy's geography was far from perfect, but unlike those who would follow, he tried to base his maps on reports from people who had actually been to the places he was drawing. While this may seem like an obvious, fundamental approach to making maps, we are assigning way too much credit to the human race.

In the centuries between 300 and 1300 CE, maps in the West were drawn less from basic observation and more from scriptural observation. The world was often depicted so that Jerusalem (or the Garden of Eden) lay at its center. This is why we talk about being "oriented": it literally means "to the east," or facing Jerusalem (from Europe). Biblical passages had a dramatic impact on mapmaking, such as a verse from II Esdras that required the land-to-water ratio to be a staggering six to one. While this would make the hurdle of reaching Asia require little more than a running leap from Spain, it would also spell the end of life as we know it, since we need water on three-quarters of our planet for things like rain to work. Yet an accurate depiction of the natural world wasn't really the goal of these maps. As the historian Daniel Boorstin put it, "they were not so much maps of knowledge as maps of Scriptural dogma."

The boy who would lead cartography back onto its observational course was raised in Antwerp twenty-five years after Columbus first landed in the Western Hemisphere. He was born as Geert (or Gerhard) de Kremer to a relatively poor family. This was a time when oats cost an incredible twenty-seven groats per *viertel*. For those who aren't experts in sixteenth-century Flemish currency, food budgets could easily consume three-quarters of a worker's total income. Fortunately for young Gerhard, his more affluent uncle sponsored him from early education to his entrance into the University of Leuven.

Of the many rules at this renowned university, one stood above them all: Thou shalt not question the ancients, and by "ancients," they meant the Bible, and Aristotle, and Ptolemy. Dissecting the story of the Creation in the Book of Genesis or Aristotle's corpus was tantamount to performing an actual dissection on an actual corpus. It was taboo and, most important, strictly forbidden.

This reverence for the ancients had created a fashion for academics to Latinize their given names. Theologian and humanist Desiderius Erasmus started life as Gerrit Gerritszoon, born to parents who obviously didn't like him very much. The famous anatomist Andreas Vesalius was born Andries van Wesel for the same reason. Gerhard de Kremer tried to reinvent himself, too. *Kremer* is the Flemish word for "peddler," so Gerhard significantly upgraded his name to Gerardus Mercator (*mercator* being Latin for "merchant," specifically the booksellers who went from town to town).

While Mercator remade his name in the tongue of the ancients, he was slowly, and quietly, losing his faith in them. He once remarked about his time in the university, "When I saw that the world of Genesis according to Moses did not fit well in many ways with Aristotle, I began to doubt the truth of all philosophers."

The authority of the ancients was beginning to crumble. Every year, explorers were bringing back information that seemed to fly in the face of previously authoritative statements. Thanks to the advent of the printing press, it was relatively easy to cross-reference new information with older sources. The inconsistencies started stacking up. Amerigo Vespucci, immortalized in the names of North and South America, had seen for himself that the claims of Aristotle and Ptolemy meant nothing when compared with actual evidence: "Let it be said in a whisper, experience is certainly worth more than theory."

Mercator was an observer at heart, a critical skill for any geographer. Being skeptical of ancient authorities such as the Bible or the *Corpus Aristotelicum* was a dangerous undertaking at the conservative University of Leuven. Yet Mercator had the spirit of a modern scientist, valuing knowledge that came through direct, reproducible observation. And thus we arrive at one of the central struggles of Mercator's life and work. Mercator was an appeaser. His sixteenth-century worldview demanded that all knowledge, both past and present, be circumscribed into one great whole. The Bible had its share of truth; Aristotle had some, too. So did other Greek, Roman, and contemporary philosophers, historians, and academics. Even the sailor and the traveling merchant and the lowly farmhand carried valuable information that, if tapped and assembled in just the right way (Go, Voltron Force!), could bring the world into the palm of your hand.

Without printing, Mercator never would have had access to the vast amounts of information that revolutionized mapmaking. In many ways, the cartographic revolution was a reaction to that flood of information. On a scale beyond any previous period of human history, Mercator's was awash with treatises and pamphlets and publications that flooded the reservoir of human knowledge. One of the great dilemmas of the sixteenth century was how the world was supposed to process this deluge. Access to copious amounts of data is one thing, but shaping that morass into a usable form is quite another. Mercator attempted to contain and redirect it into a single, clear, graceful picture of the natural world.

The ability to organize information and distribute it to the public is an incredibly powerful tool. Whether you're the Catholic Church in the 1500s or megacorporations fighting Net neutrality in the 2000s, to prioritize information is to control information. And to control information is to control people.

The fruit of the early modern world's attempts to organize all this information can be seen everywhere. One solution from the early 1600s endures today: alphabetization. Historically, information was often organized in encyclopedias and dictionaries based on subject matter. Religion would have come first, followed by law and medicine. Math and history would have been next. Frivolous works would have been branded "miscellany," or of low priority.

After the invention of the printing press, there was just too much of that information coming in. The world needed a simple, universal system to organize its data. The alphabet provided just such an egalitarian tool for dealing with human knowledge. It's why you can open a dictionary today and find *restroom* before *resurrection*, *cronut* before *crucifix*, and *goatsucker* before *god*. (It's a bird. Look it up.)

It would take thirty-five years for Mercator to produce his magnum opus, but in the intervening time, he rubbed elbows with the brightest minds of Europe. He would also be hunted and imprisoned by the Inquisition. Throughout the highs and lows of an extraordinary life, the central struggle of his work remained: how to use all that glorious new knowledge.

MERCATOR'S VENTURE into mapmaking started with a globe. Gemma Frisius (given name Jemme Reinerszoon), a member of the Leuven faculty who taught advanced mathematics, took the young man under his wing. Up to that point, Mercator had been a philosophy major. Speaking as two proud holders of humanities degrees (including one philosophy minor), we can admit that it's basically a worthless degree. (Economically speaking. Let's say.) This has been true of philosophy majors for, no kidding, five hundred years. Mercator came to the same realization, so he changed course and taught himself math. Gemma Frisius eventually brought him on as an assistant engraver for his globe of 1536, which had been chartered by Charles V, emperor of Spain.

Before print, globes were a complex, inconsistent, and extremely expensive art form. The world had to be hand-carved, detail by tedious detail, onto every paper, wooden, or metallic sphere. After print, however, an artist could create a set of plates, which would then be used to print off hundreds, or even thousands, of new worlds. You still had to paste those printed paper sections (gores) onto spherical globes, which took a considerable amount of patience and skill, but the process had become practical for mass production.

Gemma's globe was set to break new ground. According to the royal charter, it had to be "a sphere the likes of which nobody in the court of King Charles would have seen before; a globe of exquisite beauty . . ." Such a project required something far better than woodcuts as its primary medium for transferring images onto the globe. Unless you're the Da Vinci of woodcuts, they tend to appear thick and clunky on paper. (Side note: the Da Vinci of woodcuts was not Da Vinci. It was Albrecht Dürer, who exceeded the Italian artist in just this one area. Single Dürer woodcuts consistently sell for prices in the thousands of dollars, and the right woodcut, such as his famous Rhinoceros print, will leap in price into the hundreds of thousands.)

When you have to cram a whole lot of information onto a relatively tiny sphere, woodcuts just don't cut it. (Sorry for that, not sorry.) A relatively new innovation in printing was marshaled to meet this challenge: copper engraving. The difference between metal engravings and woodcuts was striking. Engraving was a sharper, more refined process. The amount of information that could be carved onto the same section of a globe was dramatically increased. In a word, the difference was resolution, and wood was the VHS of sixteenth-century printing.

But if you're going to be the Blu-ray of your era, you need a font to match. That's where Mercator came in. Over the course of a few days, he taught himself the newest fad in printing: italics. Italics (technically, *cancelleresca*, a cursive humanist script) took blocky, space-wasting Roman letters, slimmed them down, slanted them to the right, and made them sexy. Lest the reader think this an exaggeration, let it be observed that the Faculty of Theology at Leuven considered this sleek new writing style "a frivolous distraction," which is shorthand theologian-speak for anything sexy throughout the ages.

For globe making, italic fonts were much more than simple frivolity. Copper-engraved plates had already bumped up the level of resolution in printing, and italics provided a compressed, precise, visually appealing script to complement that.

From his involvement in this project, Mercator would also have learned firsthand the value of eyewitness reports in cartography. There was another collaborator on Gemma's globe, an older gentleman and lifelong diplomat named Maximilianus Transylvanus (originally Max von Sevenborgen). Transylvanus was best known for writing and publishing the first report of Magellan's circumnavigation of the planet. He was basically brought onto the globe project as a fact-checker. Among other things, he completely discredited the existence of Taprobane, the world's largest fake island, which had been cited by geographers as far back as Ptolemy. That's like finding out Australia has been a lie this entire time. (Although, after being introduced to Vegemite, you might be willing to believe the whole continent is an elaborate hoax to get you to put something that tastes like koala scat into your mouth.)

Despite hitting delays and setbacks, Gemma finally completed his globe, in all its glory, in 1536. The success of that endeavor pushed Mercator to the next logical step, known in the sixteenth century as the *Be-yon-cé*: leaving his crew behind and going solo to create some of the most triumphant works of his time. For the subject of this cartographical debut, Mercator hedged his bets and started with the one place that had been mapped in the West more than any other: Europe. Just kidding, it was Palestine. Because the Bible.

The first Bibles with maps had arrived in Antwerp about a decade earlier. Affixing real-world coordinates to Bible stories such as the Exodus helped validate them in people's minds. Mercator's Palestine map was the best of its kind, and considered the gold standard of biblical mapmaking for five decades after its first printing. The standard before Mercator was to read the Bible, point at locations on a map, and say, *This looks like a fine place for Sihon, King of the Amorites, who dwelt in Heshbon and ruled from Aroer. Hell, why not? I literally know what none of those words mean.*

Besides actually checking his sources, Mercator also made a note when he was just guessing, thereby drawing a clear distinction between making shit up and making educated shit up. He was still rooted in a belief that the Bible and the natural sciences could be brought into harmony. It took time for him to let go of this inherent trust, even in the face of contrary evidence. As his maps progressed, however, this Palestine-centric view of the world gradually receded. When his magnum opus was published thirty years later, Jerusalem would be barely visible, the birthplace of Christianity reduced to a mere abbreviation among all the other place names.

Mercator's 1537 Palestine put him on the map. (Sorry. Again.) Being the ambitious twenty-five-year-old he was, he next set his sights on the entire world. The map he created in 1538 would be the most detailed of its time. And by "time," we mean two years. New information was being printed like a steady stream of machine-gun fire, shooting to pieces any maps that were more than a couple of years old. Mercator's 1538 world map was based on Gemma's 1536 globe, and both would be obsolete by 1540. As much as people find the sprint of modern technology absurd at times (a new iPhone every year?), the Age of Discovery saw a similar uptick in obsolescence. Maps became little more than decorative wall paneling after just a few years. (Not that there's anything wrong with original Mercator wall paneling. We'd take that in a heartbeat.)

The consequences of the Age of Discovery had much in common with those of the invention of print: both provided unprecedented access to knowledge, and therefore both were subject to political and religious freak-outs. It was science versus authority. What happens when power and evidence collide?

Make no mistake, knowledge is power, and accurate maps had potentially devastating effects on the balance of power in Europe. Maps became "instruments of rule" and cartography "the science of princes." Trade routes could make or break economies. For a Spanish or Portuguese sailor, sharing maps or charts outside the country was punishable by death. Publication of the maps of Francis Drake's circumnavigation was initially prohibited. Even printed accounts of famous voyages could be heavily edited for state security, as the printer for the narrative of Frobisher's voyages admitted in 1578: "I have in a few places somewhat altered from my copy, and wronged thereby the author, and have sought to conceal upon good causes some secrets not fit to be published or revealed to the world (as the degrees of longitude and latitude, the distance, and true position of places, and the variation of the compass)."

Military invasions could be planned, down to the most minute detail, from the comfort of one's sitting room. As the moral backbone of the twentieth century has taught us, "Remember kids, knowing is half the battle." Spies were dispatched across Europe specifically to bring home maps. Models of French eastern border towns were classified as state secrets well into the twentieth century. The entire continent of Africa was treated as a state secret by Portugal, which would have been a pretty impressive feat had it worked. Even a commentary on Roman architecture was considered too dangerous to print in Venice because its depictions of fortifications could have helped rival states undermine its defenses.

In the sixteenth century, there were real-life consequences for anyone who drew, printed, or distributed maps. Mercator learned this firsthand in 1538, when he was roped into the power struggles of the Low Countries. The Flemish province of Ghent (in modern-day Belgium) had been in open rebellion to Emperor Charles V and his sister, Queen Mary of Hungary, since 1537. The flash point for this uprising was taxes. Of course.

In that same year, Queen Mary sent tax collectors to help finance her brother's military campaigns to kick the *merde* out of godless France. Ghent's guildsmen (your weavers, glaziers, locksmiths, jewelers, printers, and the like) thought the endeavor a waste of their money, so they took up arms and seized the gates. Because Charles V was off in Spain, too obsessed with force-feeding Protestant François I his own *mille-feuille*, the city of Ghent fantastically achieved a bloodless coup against the head of the Holy Roman Empire.

And what better way to celebrate independence than a huge block party? Just such a citywide festival was conducted over the spring and early summer of 1539. Poetry and theater and rousing oratories were performed that lambasted the Catholic Church and patted Ghent on the back for a revolution well done. Things did get a teensy bit out of hand when a seventy-five-year-old former politician was seized and forced to shave his head and beard—basic drunken party antics. Then the old man was ritually tortured to death on the rack; apparently the shaving was done to ensure that demons couldn't hide in his hair and help ease the pain. When other former aldermen met similar fates, panic spread, and the upper crusts of Ghent fled the city.

Prior to the festival, an important part of the province's declaration of a free Ghent had been the commission and printing of a new map of the Netherlands, one that depicted an independent and proud Flanders. With a few pen strokes, even eastern Flanders was swept up into the revolution with the label "Under Ghent." The citizens of cities such as Leuven (where Mercator was living) and Antwerp were probably thinking, *For the love of God, leave us out of this*, but that was the political power inherent in mapmaking. No longer did you have to dust off your *Zweihänder* and march out to conquer a domain. You could just grab a printing press, draw some dotted lines, and proverbially call, "Shotgun!" on whatever you saw. (All levity aside, that *is* how maps were made in Europe for hundreds of years: powerful white men carving up the world with little regard for the potential fallout of their cartographical doodles. That never blows up in anyone's face, as we know from indigenous peoples everywhere and the entire Middle East.)

Emperor Charles V, however, wasn't about to honor squatters' rights, even if they did have a nicely printed map to prove their claim. He gathered his army and set out from Spain to literally and figuratively tear that map to shreds. Suddenly realizing the lethal situation in which they'd placed themselves, the guildsmen of Ghent ran back to their printing presses to make restitution. They hired Mercator and gave him three months to fix the inflammatory map of Flanders, hoping this would appease the wrath of the Holy Roman Emperor. By all accounts, Mercator was happy to do it. Preventing the obliteration of a province in Flanders is a win-win for everyone in the Low Countries, especially when that province stupidly labeled your house "Under Ghent."

Mercator's 1540 map of Flanders was magnificent, not only in its technical advances—he worked with a surveyor who practiced a newfangled technique called triangulation, making it "the first truly accurate printed map in the Low Countries"—but also in its distinguished levels of back-pedaling and imperial ass-kissing. Borders were redrawn, banners from proud Flemish cities thrown out, and rebel strongholds erased, all to reinforce the idea that Flanders was an obedient imperial realm. The edges on the map of this once acerbically independent country were even ornamented with medallions and coats of arms that traced the genealogy of its royal succession, demonstrating clearly that Emperor Charles V was the rightful ruler. Slap a frilly dedication to the emperor at the bottom, and (brush the dust from your lapel): crisis averted.

It did not work. That bottle had been uncorked, and some things just can't be unsaid . . . or redrawn . . . or unexecuted on the rack. Charles, with his sister, Mary, crushed Ghent using an army that took five hours just to march through the front gates. Thirteen instigators of the rebellion were ceremoniously beheaded, and droves of Gentenaar were draped in black and forced to walk through the city barefoot to beg the Holy Emperor's forgiveness. Whoever said the pen is mightier than the sword—it was English novelist Edward Bulwer-Lytton—Ghent has a middle finger with your name on it. In the end, Charles won his colony back, and he even got a new map to prove it.

Mercator escaped from the Ghent debacle fairly unscathed. In fact, it might even have been a net gain for him. After completing a globe of his own in 1541 (which made his 1538 world map look like a shitty iPhone 4), Mercator received a requisition for globes and geographical instruments from none other than Charles V. Mercator happily agreed. Having an emperor as a personal client in the sixteenth century was a ticket to stardom—which makes the next chapter of Mercator's life all the more inexplicable.

IN 1543, Charles's sister, Mary, along with the Inquisition, launched a new invasion of the Low Countries. This time they were ferreting out Lutherans and helping them see the error of their ways through the transformative power of Christ's love. And also torture. Like Tyndale before him, Mercator 's name ended up on a list of forty-three in the Low Countries sought as possible heretics.

How Mercator's name got on that list is a bit of a mystery. The list mentioned *litteres suspectes*, "suspect letters," as the only accusation against him. Officially he was charged with *lutherye* and evading arrest. Because Mercator was careful about what he said and did in public, any Lutheran leanings on his part would have been difficult to prove in court. The violent conflicts that accompanied the Reformation had been raging most of his life, and he knew how to keep his head down. The charge of evading arrest, however, appeared a little clearer cut.

In February 1544, officers of the Inquisition showed up on Mercator's doorstep, but the famous cartographer wasn't there. His uncle (the one who had overseen his education) had died and Mercator had traveled just that morning to his hometown of Rupelmonde to sort out the details of the inheritance. Your beloved uncle dies; that's bad. He leaves you some inheritance; that's . . . good? The Inquisition shows up at your door, doesn't believe the coincidence, and declares you *fugere*; that's bad. When it comes to the Inquisition, "going on the lam" and "entering a flagrant guilty plea" are the exact same thing.

Mercator was arrested in Rupelmonde and imprisoned in the castle there for seven months. During that time, he would have been in isolation and allowed no contact with the outside world. Seven months of solitary confinement is enough to break anyone's soul, but Mercator wouldn't even have known the charges against him, or have been informed of the efforts that his friends and family were making to secure his release. It would have been one long, hopeless night of despair.

Since Mercator "ran from the law" to his uncle's house, he was presumed guilty, but even the Inquisition required some kind of actual evidence to execute him. Three months after his arrest, secret investigators were sent to the monastery of Franciscus Monachus, a monk and fellow cartographer who openly criticized ancient authorities such as Aristotle and Ptolemy (his actual name was Frans Smunck, but obviously he couldn't be known as Smunck the Monk). Somehow, the investigators had come to the conclusion that incriminating letters could be discovered at the monastery that would implicate Mercator as a Lutheran. If there was any truth to the allegation—Mercator and Monachus had been friends and pen pals for years—those letters didn't exist any longer, and the investigators came back empty-handed.

The Inquisition was one of the best record-keeping institutions of its time. Historian John Arnold writes that their records "permit[ed] the operations of various kinds of power: catching out suspects, permitting future investigations, constructing future transgression." Luckily for Mercator, that investigation faltered. In the margins of one document that contained his name and a reference to suspicious Franciscan letters, "no h" has been hastily scribbled. Presumably this means *non habent* ("they do not have them"). The evidence of

Mercator's treason was therefore lacking, and four months later he was finally released, after seven months' imprisonment.

If the members of the Inquisition had found what they were looking for, Mercator's greatest achievements would have been annihilated twenty years before going to print—or, more accurately, strangled, disemboweled, and then immolated twenty years before going to print. The Inquisition sought to obliterate not only the publications of heretics, but the heretics themselves. Of the people arrested at the same time as Mercator, at least two were buried alive, two were burned at the stake, and one was beheaded.

Considering the black comedy of errors that had to happen for Mercator to land in Rupelmonde Castle, the guards could have slapped him on the back and said, *You'll laugh about this later.* But no, he wouldn't. One of the conditions of his release was that he never speak about the case again. As far as we have on record, he stuck to that. In a letter to his friend Antoine Perronet, he wrote only that his time in Rupelmonde was "the most unjust persecution." That was it. (Mostly, anyway; near the end of his life he *would* express his belief that the Catholic Church was the horned beast with iron teeth that ends the world in the Book of Daniel, but after seven months of false imprisonment in a castle dungeon, we'll applaud him for his restraint.)

One of the great ironies of this wrongful imprisonment is that, at the time he was arrested for treason against Charles V, Mercator was under contract to produce surveying instruments for . . . Charles V. These instruments were being used to wage war against the very Lutherans with whom Mercator was accused of sympathizing. The instruments were so vital to Charles's war efforts, in fact, that after getting melted in an artillery raid in Germany, the emperor would personally write Mercator to ask for more. Ten years later, Mercator would even be named *Imperitus Domesticus* ("member of the imperial household") as a gesture of appreciation and, one hopes, a *sorry for the whole business with the dungeon and the almost burning you alive. Our bad.*

OVER THE next fifteen years, Mercator moved to the now-more-temperate political climate of Germany and worked on his maps, globes, and surveying instruments in relative quiet. When he wasn't out in the field triangulating the planet, he was coordinating the printing of maps with his sons, who worked alongside him. Yep, things were pretty idyllic—except for one nagging issue that had been on the back burner of Mercator's mind for more than thirty years.

Back in 1538, Mercator had tried to map the world. When you do this, you inevitably run into a thorny issue. How do you accurately represent the surface of a spherical object on a flat piece of paper? How do you take something 3D and make it 2D? Mathematically speaking, this comes with a whole mess of problems. You can make sure all the latitudes and longitudes match up, which forces the sphere's land masses to flatten and distort. Or you can preserve the proportions of the land masses, but lose the technical accuracy of the latitudes and longitudes. There is no easy answer. Making a sphere into a sheet will always create distortions.

Creating a world map that enabled a sailor to draw a course and then follow that course on a round world was not commonly known before Mercator. Straightening a globe out and printing it on a piece of paper is serviceable for small journeys, such as crossing the Mediterranean, but in the sixteenth century, people were circumnavigating the planet. At such great distances, small miscalculations could put a navigator hundreds of miles from his intended destination.

Mercator's world map of 1538 attempted to address this issue, known as a projection, by making the map image heart-shaped. It was a striking piece of work, but it failed miserably. Not only could you not plot a straight course to your destination, but the heart shape warped and distorted the continents unnecessarily. Beautiful yet useless.

Thirty years later, Mercator had a breakthrough. It was a simple breakthrough, but a profound advancement for modern cartography. On a globe, the spaces between latitudes shrink the closer you get to the poles. If you were to cover a balloon with vertical lines, you'd see that the lines get more and more crammed as they reach the bottom and top of the balloon. Mercator realized that if you simply straightened everything out in equal proportions in all directions, you could maintain a flat, gridlike map that didn't lose its global integrity. This revolutionary technique would later be dubbed the Mercator Projection.

When discussing the fallibility of scientific discovery, some people like to trot out the old "people in Columbus's day thought the world was flat" story. Not only is that just plain wrong—Columbus thought he could sail to India precisely *because* he knew the world to be round—but mathematicians for thousands of years before Columbus knew that their world was a sphere, and they attempted to map it accordingly. In fact, Columbus used Ptolemy's calculations of the size of the earth to demonstrate how short he believed the voyage to be. It wasn't until Mercator and his projection that the human race finally figured out how to make their round world flat.

Stretching everything out equally allows a sailor to accurately plot a course, but there is a drawback: that equal spacing also means that the land masses near the poles are going to have to stretch in order to compensate. How far might they have to stretch, you ask? Well, infinitely.

You may remember seeing a Mercator Projection map like the one on the following page in pretty much every elementary school classroom you've ever entered.

Land masses at the equator are relatively normal-size, but the

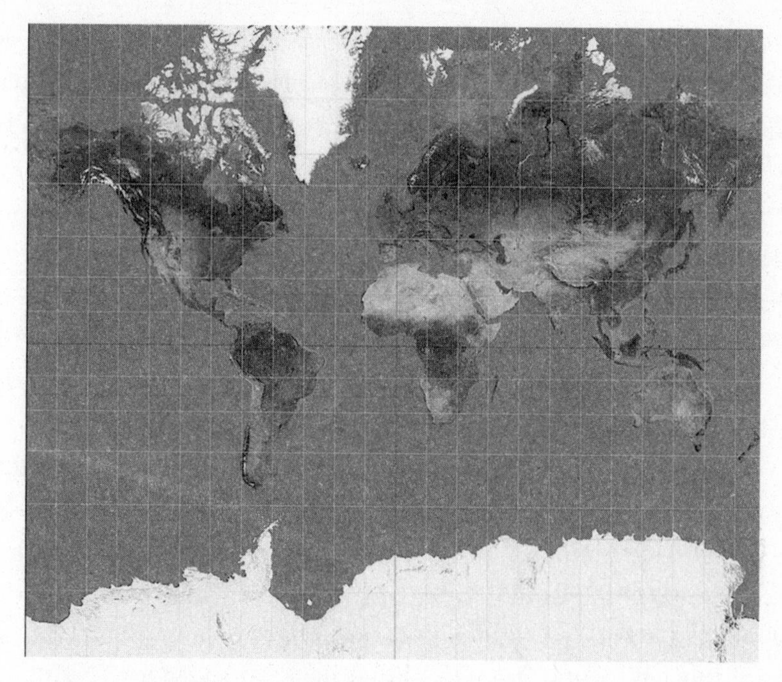

THE MERCATOR PROJECTION, USED IN A 2011 MAP OF THE WORLD.
Courtesy Daniel R. Strebe.

farther north and south you go, the more they stretch. Greenland is a large island, but it certainly isn't larger than South America. In reality Greenland is about two million square miles *smaller* than Australia. Eight Greenlands could fit into South America, with about two hundred thousand square miles to spare. No matter what those elementary school maps tell you, Alaska is not that big. Neither is Canada or Russia. And Antarctica? That's what it looks like when you stretch a continent to infinity.

As we've seen with other accidents of misinformation, these distortions can have surprisingly far-reaching consequences. In the twentieth century, the Mercator Projection was a favorite of the anticommunist John Birch Society, which would use copies of the projection to illustrate its lectures. The enormously inflated areas of Russia and China created a deep red wave that seemed to bleed across the whole world. Maps are extremely effective propaganda tools because they are both visually impactful and terrifically easy to manipulate.

Still, no matter what kind of Violet Beauregarde Wonkaesque treatment you give those continents, their positions on the grid are 100 percent accurate to real life. You can plot a course from Tunisia to Chile using that image, and mathematically, the latitudes and longitudes will all work out. That is the simple and somewhat counterintuitive brilliance of the Mercator Projection.

Mercator's revolutionary map of 1569 is not only his magnum opus, but the very embodiment of practical mapmaking (well, except for the Arctic insert with its nonexistent islands of healthy, happy pygmies). The map is a sight to behold. Gone are the fancy symbolic designs and cartouches of Middle Age maps. The biblical decorations disappear, and the birthplace of Christianity is reduced to an abbreviation. True, there are the occasional flourishes, such as the painfully racist drawings of cannibals from South America preparing a human smorgasbord (nobody's perfect), but the purpose of the 1569 map was practicality. Unlike most sixteenth-century mapmakers, who made up coastlines as they went along, Mercator meticulously checked and rechecked his sources, built on the knowledge that had come before, and worked tirelessly to marry the old data to the new.

He could do this only because of the printing press, which allowed him access to sources on a scale that had been practically impossible before. This makes his maps some of the most dazzling historical artifacts from the dawn of modern science.

Unfortunately, since sailors initially didn't know how to use the new projection—this was a time when directions for sailing from Europe to the East Indies included sailing "south until the butter melts, then due west into the sunset"—Mercator's literally earth-changing map gathered figurative dust for decades after the mapmaker's death. Not until 1599 did an English mathematician, Edward Wright, explain the projection and publish charts to counter the inevitable land mass distortions.

In modern times, this projection holds a place of cartographic royalty. It was officially used to survey both England and America in the twentieth century. Mercator is even now contributing to the space age with Mars, Venus, Jupiter, Io, and Titan each mapped with his projection. Watch out, Red Planet, the Sino-Russian distortion is coming for you.

As forward-thinking as Mercator was, however, he still fell victim to one of the ailments of any information age. In addition to the interesting but totally wrong curiosities of the Arctic Circle, chunks of misinformation made their way onto his 1569 map. For example, he dropped the mythical Christian kingdom of Prester John right into the middle of Africa. And why not? Africa is goddamn huge. Who is going to prove that an imaginary Christian nation hasn't been hiding out in Ethiopia for fifteen hundred years? (The answer is Portuguese scholar Manoel de Almeida, about a hundred years later.)

Whether it's Mercator's day or our own, it's important to remember that not all information is created equal. We would do well to remember the "world is flat" argument from Columbus's day. Scholars in the fifteenth century certainly didn't believe it, but did *anyone* at the time think the earth was flat? Yes. Absolutely. But they were the uneducated, the bumpkins, the lummoxes, yokels, and yahoos. Imagine if, five hundred years from now, people judged the collective knowledge of the twenty-first century on Internet comments from . . . well, Yahoo. Some of our favorite scientifically motivated questions on Yahoo Answers include: "Do you think NASA invented thunderstorms to cover up the sound of space battles?"; "What percentage of water is celery?"; and "How big is the specific ocean?" To people from the year 2517: for God's sake, please remember us for our best instead of our yahoos.

As mistakes like Prester John or the Arctic maps prove, even brilliant geographers such as Mercator had their blind spots. Unfortunately, the ability to print something (or post something on the Internet) does not in and of itself guarantee veracity. One of the biggest challenges of the Internet age is how to examine sources critically. Just like us, the sixteenth-century head-butted this issue repeatedly. It's easy to look back from a modern perspective and see the weaknesses of the great men and women who came before. But when it came to organizing the world's printed information and misinformation, Mercator the mapmaker stands out as one of our less gullible heroes.

CHAPTER FIVE 5

BAD SHAKESPEARE

AFTER LEARNING OF HIS FATHER'S early death, Prince Hamlet of Denmark wallows in despair. He contemplates ending his own life, and from those pain-racked lips falls one of the most quoted monologues ever uttered:

> *To be, or not to be; that is the question:*
> *Whether 'tis nobler in the mind to suffer*
> *The slings and arrows of outrageous fortune,*
> *Or to take arms against a sea of troubles,*
> *And, by opposing, end them. To die, to sleep—*
> *No more.*

From the mind of Shakespeare, to his pen, to the words before you, Hamlet's soliloquy is among the finest ever crafted by the great Bard. Or was it? There is another version of Shakespeare's *Hamlet*, the earliest printed version, that is somewhat less refined in the philosophizing of the crown prince. "To be, or not to be, Aye there's the point, / To Die, to sleepe, is that all? Aye all." These yokelish lines belong to a version of Shakespeare's *Hamlet* that, for a hundred years, scholars called the "bad" quarto.

Single plays were most often printed in quarto format, meaning four pages to each side of a sheet of paper. In Elizabethan England, these quartos were roughly the dimensions of a cheap square paperback book. What made some of these particular quartos "bad" was how rough they were in comparison to later versions of Shakespeare's plays. It's like when someone says this is a "bad casserole." You do not eat that casserole. It will make you puke your guts out. For almost two centuries, scholars felt the same way about "bad" *Hamlet.*

But what if the "bad quarto" isn't really that bad? What if that quarto is just an earlier version of *Hamlet*? Or what if Shakespeare, arguably the most important writer in English literature, wasn't really as good as we remember him today? And what if the Bard's reputation was shaped, in part, by the people who memorialized him in print?

For two hundred years after Shakespeare's death, scholars had no idea that an earlier version of *Hamlet* even existed. Then, in 1823, a man with the very English-sounding name of Sir Henry Bunbury stumbled across a copy at his Barton Hall estate, in Suffolk. He would later document this amazing find as a footnote to a memoir he was writing about someone else: "the edition of 1603, the only copy of which, known to be in existence, was found by me in a closet, 1823." Bunbury writes this so nonchalantly that it would appear as if English closets were routinely the sites of astonishing discoveries. (Behind this closet door, a land of mythical creatures led by a talking lion. Behind this one, we keep a boy wizard. Behind that door, a previously unknown work of Shakespeare that will turn the literary world on its head.) Sir Henry Bunbury sold the collection to the booksellers Payne and Foss for £180; they quickly turned around and sold it "at a tidy profit" to a friend of Charles Dickens.

Thirty years later, a second "bad quarto" was discovered—likely hidden for so long because it was missing its title page—and sold to a bookstore in Dublin. Shockingly, the bookseller, M. W. Rooney, had a hard time selling this book at first. Since it was an incomplete copy, he was ignored by the British Museum, which considered his asking price too high. Yet this was literally one of the only known copies of the earliest *Hamlet*! Some of us would saw off our own pinky fingers just for a chance to hold the thing. Sure enough, a presumably repentant British Museum did end up purchasing the quarto through a private collector (for more than Rooney had initially offered it).

These two copies are all that have survived of the earliest-known printing of Shakespeare's *Hamlet*. The discovery of the 1603 *Hamlet* and several other "bad quartos" (including such well-known plays as *Henry V*, *The Merry Wives of Windsor*, and yes, even *Romeo and Juliet*) forced the world to face the very real possibility that Shakespeare may not have been as great as we remember him. Some scholars have spent their entire lives debating these points, fighting among themselves to explain how Shakespeare became Shakespeare.

To descend into the world of Shakespearean scholarship is to descend into a particular species of madness. Factions with names such as the Disintegrationists, the New Bibliographers, and the Revisionists rise like Elizabethan houses to duel one another with collating machines and proof sheets and watermark catalogues. They lob insults at one another: "Bardolator" (bard + idolatry) and "Bardoclast" (bard + iconoclasm). Diving into the nerdy carnage in their wake tests the mettle of any researcher, and at a certain point, you'd rather take a poisoned rapier to the heart than read one more goddamn textual criticism. But if you take anything away from the labyrinth of that scholarship, it should be this: what you think you know about Shakespeare may not be so.

Was Shakespeare the best-selling playwright of his time? Yes. Could just putting his name on a title page sell books, even if they weren't his? Absolutely. Was Shakespeare an insightful storyteller whose writing ranged from the sublime ("Duke Orsino: If music be the food of love, play on . . .") to your garden-variety smut jokes ("Chiron: Thou hast undone our mother. / Aaron: Villain, I have done thy mother")? No doubt. But when we look at Shakespeare's plays today, we simply cannot ignore that, on some level, centuries of editing have fine-tuned and honed what we know as the Works of William Shakespeare. The man who wrote these works was a real person whose fallibility and roughness has been smoothed out over time. Yet, stripping away that polished veneer is a worthwhile endeavor. Only by discarding the dust of our reverence do we get a clearer picture of the brilliant writer remembered by the world as William Shakespeare.

Now, not all scholars have agreed on this point. The struggle to keep Shakespeare on his pedestal has influenced how editors have presented his plays to generations of readers who were happily ignorant of the warfare raging behind the scenes. As scholar R. B. McKerrow summarized in 1933, "if an editor likes a reading, that reading is (a) good, and (b) attributable to Shakespeare."

It was more than just editors, though. Books are not a direct line from the minds of authors to their readers. Many people along the way have their hands in that cookie jar, and Shakespeare was no exception. Publishers, printers, typesetters, and even the actors and playhouses before them—all had an effect on Shakespeare's plays. Almost all the changes they made to his work occurred without Shakespeare's participation or after he was already dead. So how do we determine what an "authentic" Shakespeare play would have looked like?

Let's take a minute and consider the famous eighteenth-century poet Alexander Pope. In the 1720s, Pope edited an edition of Shakespeare's plays that paid special attention to the earliest printed texts, following a "historically based editorial practice." For instance, he took pains to compare different editions of the same play. This seems obvious today, but in that period it was a notable change in editorial philosophy. As an example, one of the reasons the fourth collected edition of Shakespeare's plays (the Fourth Folio) survives more than any other is due to seventeenth-century buyers assuming it was the most up-to-date and therefore the most accurate. This mind-set prompted folks who owned earlier editions to start tossing them out after they had purchased the new one . . . a physically painful realization for any historian or collector. To twist the knife even further, the first collected edition of Shakespeare, the First Folio, today sells for between $4 million and $6 million; copies of the Fourth Folio sell for around $200,000 to $250,000.

Pope may have been the first to look back to the earliest printed Shakespeare texts, but even he wasn't above tweaking the Bard. He would revise Shakespeare's verse when it seemed to show metrical errors, and occasionally "update" the text for contemporary readers, removing verses or wording he didn't like. He moved about fifteen hundred "degraded" Shakespeare lines to the footnotes, when he kept them at all.

Pope's edition was viciously attacked by the scholar and translator Lewis Theobald in a 1726 work called *Shakespeare Restored: or, a Specimen of the many Errors, as well committed, as unamended, by Mr. Pope in his late Edition of this Poet. Designed not only to correct the said Edition, but to restore the True Reading of Shakespeare in all the Editions ever yet publish'd.* The criticisms in this snappy little title range from petty interpretations to major misunderstandings of Shakespeare's work. Mostly, Theobald took issue with Pope "refining" Shakespeare's style into what was trendy in the eighteenth century. In response, Pope made Theobald the dull, maligned chief of the dunces in his new poem, *The Dunciad.* This attack in verse is one of the crowning achievements of an era celebrated for its satirical bitterness. One scholar calls *The Dunciad* "the greatest work in

English literature to which Shakespearean controversy has given birth."

Writing one of the great works of verse of your era, however, doesn't save you from legitimate criticism. After the dismal sales of Pope's Shakespeare, his publisher turned traitor and chose Theobald, of all people, to edit their next edition of the Bard. Ouch. Of course Theobald had his own issues with Shakespeare. "There are very few pages in Shakespeare," he wrote, "upon which some suspicions of depravity do not reasonably arise." This cycle of suspicion is probably the single unifying link between the major editors of Shakespeare across hundreds of years. They all agree that the play texts are suspect, even if they can't agree on anything else.

Surprisingly, the one person who appears to have been the most blasé about the interpretations of his plays is the man himself. Outside of possible rewrites (one of the many proffered explanations for the existence of earlier "bad quartos"), there is no direct evidence that Shakespeare was concerned with how his plays would be remembered.

This wouldn't have been out of place for the time. Plays were usually sold to a theater team at a price of around six to eight pounds. Shakespeare worked with the Chamberlain's Men, which became the King's Men in 1603. He could sometimes offer revisions to his plays, but the troupe itself was free to make changes to the text as they saw fit. It's a bit like authors selling their book rights to a film production company. Once sold off, the adaptation belongs to the company. It can do whatever it wants with it. It can make it way better than the original (as in the case of Francis Ford Coppola's *The Godfather*), or it can murder it and bury it quietly in the backyard (like James Franco's 2013 *As I Lay Dying*).

Shakespeare wrote plays for the same company in which he was acting, so he likely retained some influence on the texts. Yet other people, including very powerful Elizabethan figures, made their influence known as well. After leaving Shakespeare's pen, his work would have been perused by the Master of the Revels. If the title isn't obvious enough, this actual member of the royal household was in charge of all royal festivities. He was also in charge of censoring plays to make sure "nothing too seditious or blasphemous was played on the stage." Luckily, both for English audiences and for every eighth-grade literature class, thinly veiled references to lady parts were considered neither seditious nor blasphemous. (Reading a letter from his employer in *The Twelfth Night*, Malvolio says, "By my life, this is my lady's / hand. These be her very c's, her u's, and ['n'] her t's.") At a time when the word *nothing* was a euphemism for vagina (no-thing), *Much Ado About Nothing* isn't even trying anymore.

After the Master of the Revels granted his approval, a play could be altered to fit the needs of a particular theatrical troupe or performance. With its focus on action and plain language, one past theory of the *Hamlet* "bad quarto" suggests the play is an abridged traveling version used by the King's Men.

As the play texts passed from actors to private investors to publishers and printers, changes were unavoidably made along the way. One particularly grievous theory of the "bad quartos" involves memorial reconstruction. Rather than making changes to a play text in front of him, memorial reconstruction involves an actor from a troupe, one with a bit part, reconstructing the play from memory, writing it all down, and selling it to a publisher. Thomas Heywood, a poet and playwright contemporary to Shakespeare, observed, "some of my plays have (unknown to me, and without any of my direction) accidentally come into the printer's hands, and, [have become] therefore, so corrupt and mangled (copied only by ear)." However, it is questionable just how widespread memorial reconstruction was in Elizabethan England for drama.

The only instance of Shakespeare seemingly exhibiting displeasure comes to us secondhand, from the previously mentioned Thomas

Heywood, but it wasn't even about his plays. In 1599 a printer named William Jaggard published a collection of poems entitled *The Passionate Pilgrim*, attributing the entire thing to "W. Shakespeare." As it turns out, only five short poems in the 120-page octavo had actually been penned by Shakespeare, and the Bard wasn't too happy about this little advertising scheme. "The author [Shakespeare] I know [was] much offended with [W.] Jaggard (that altogether unknown to him) presumed to make so bold with his name."

But here's the thing: it wasn't just Jaggard. Shakespeare's name was big business, even while he was alive and kicking. No other playwright with the initials W.S. "was deemed worthy of publication" between 1590 and 1616, yet a number of play texts found their way to market with just those initials. Shakespeare scholar Lukas Erne borrows from *Romeo and Juliet* to make the point "What's in a name? . . . money. [And] a name to make money with was 'Shakespeare.'"

Scholars have been inclined to believe that disapproval from Shakespeare worked when it came to *The Passionate Pilgrim* because in the 1612 edition, Jaggard removed Shakespeare's name from the title page. This dispute with Jaggard concerned Shakespeare's poems, however, a much more respected literary medium at the time. When it came to his plays, we have no documentation for how Shakespeare felt.

It might best suit our purposes to avoid the black hole of time and hope that is Shakespearean textual criticism and focus instead on how the history of print influenced what we know of the famed playwright. That Shakespeare is one of the most important writers who ever lived is not exaggeration. That his plays have had a profound impact on four hundred years of Western civilization is beyond question. All this notwithstanding, it's just a fact of history that we still don't "know Shakespeare" (a little tip of the hat to the master of the double entendre). "Any tale that scholars tell about these plays must on some level be a story about how little we know, or our story will not be true," observes Shakespeare scholar James J. Marino.

We might start with Shakespeare's name. There are eighty-three variants. Not too surprising in a time before "the dictionary" or "standard spelling." More important, we have six autographs that have been directly attributed to the Bard. The spellings range from "Shakspeare," to "Shakspere" to even "Shakspe" and "Shak sp." The common thread here is the notable absence of the letter *e* after the *k* in *any* of his signatures. So how did "Shakspeare," which our spellcheck is highlighting angrily, become "Shakespeare," which irritates exactly no spellchecks? The answer is the printing press.

When typesetting Shakespeare's name, specifically in italics, the *k* and the antiquated long *s* (*ſ*) overlapped. Under the mechanical pressures of the printing press, the two letters tended to chip or break. In order to resolve this issue, compositors slapped an *e* between the letters, a typesetting practice called kerning. "Shak-" became "Shake-," the *k*'s and *ſ*'s were saved, and the great Bard was condemned to have his name misspelled by everyone everywhere for the rest of time.

Both the inevitable processes of printing and the equally inevitable mistakes of compositors have had an impact on how we read Shakespeare, and we don't even realize it. As book historian Roger Stoddard famously put it, "Whatever they may do, authors do not write books. Books are not written at all. They are manufactured." This process of manufacturing leaves its own marks.

When planning the printing of a particular volume, printers had to estimate just how much text would fit on each page, a practice called casting off. If these calculations weren't accurate, the compositor was faced with a real problem, since he couldn't just hit Backspace and magically reformat the document. In these cases compositors might cram more text onto the page, or simply cut lines.

There's even the (remote) possibility that some lines were added to plays to pad out a block of text that was too short. As one scholar put it, "The worst-case scenario is that the compositor might feel compelled to add the odd word, phrase, or clause to fill out a speech and get it into a new line. The prospect of Shakespeare's quartos containing material 'written' by a compositor trying to fill out a page fills bibliographers with horror."

While the practicalities of casting off wouldn't have had a huge impact on our interpretation of Shakespeare, typesetting mistakes certainly have. In *Richard II*, Sir Stephen Scroop approaches King Richard to inform him of how deep the rebellion against him runs. "White beares have armed their thin and hairless scalps against thy majestie." Apparently King Richard was so despised that even the follically challenged wild animals of England were reaching for their swords—or the typesetter's hand slipped into the *e* box of type, which sat next to the box of *d*'s After all, "White beards [old men] have armed their thin and hairless scalps against you" makes a whole lot more sense.

For years, people were confused about a seemingly nonsensical list of questions found in a speech in *The Merchant of Venice*—until it was realized that the compositor had just run out of periods and substituted question marks in their place. That doesn't change the meaning at all, does it./?

In the middle of *King Lear*, "Edmond," an illegitimate son of the Earl of Gloucester, gets a name change to "Bastard" in the stage directions and speech prefixes. Entire studies have been written on the significance of this appellation, as if the change showed that "his 'bastard' birth shaped and defined Edmond's true self." Or it's entirely possible that the capital *E*, which was in heavy demand in a play text with frequent *Enter*s and *Exit*s, was sidestepped by calling [E]dmond "the Bastard," a move that had no significance whatsoever outside the printing shop.

So the history of print mangled Shakespeare a bit here and there, but it made up for it by immortalizing him to the ages. Single quartos were printed sporadically throughout his career as a playwright, but the first attempt to gather his plays into a printed "collection" of great Elizabethan dramas took place in 1619, three years after his death. And who better to print those plays than the object of Shakespeare's one recorded resentment: William Jaggard.

The printer of the falsely attributed *Passionate Pilgrim* was doing rather well for himself. In 1608 he purchased the business holdings of James Roberts, a well-respected London printer who had already published a few Shakespeare plays. By 1610, Jaggard was named the official Printer to the City of London, the nexus of which was the storefronts that lined St. Paul's Cathedral.

At first glance it might seem odd that a cathedral churchyard could be the center of the Elizabethan book trade, but the architect of this bizarre state of affairs was the same as many other bizarre states of affairs in sixteenth-century England: Henry VIII. In 1534, Henry took upon himself the title of Supreme Head of the Church of England, effectively gut-punching the authority of the Catholic Church. This was followed two years later by a wave of legal actions that dissolved every monastery, convent, and friary in England, Wales, and Ireland. All income and assets were pillaged and turned over to the Crown.

St. Paul's Cathedral, built five hundred years before, was appropriated and turned into a center of commerce. Overlooking the irony of Jesus casting out the moneylenders from the temple (or perhaps a blasphemous nod to it), the Catholic chapels, shrines, crypts, and other holy buildings in the churchyard were sold off as rental spaces to entrepreneurs, who were mostly Protestant. At one point, fish, fruit, ale, beer, and "other gross wares" were being sold in St. Paul's, so in comparison, selling books there would not have been that odd. Books are at least two times more dignified than fish and six times more dignified than beer.

By 1611, St. Paul's Churchyard was the respectable heart of the London book trade, and William Jaggard was the man to whom you waved as he sauntered by your wares. It's unlikely Jaggard would have waved back, however. By 1612, William Jaggard, the Printer to the City of London, was going blind.

Ah, syphilis. In a time before penicillin, when the primary cure for syphilis was highly toxic mercury treatments, we don't know if Jaggard went blind from his STD or its "cure." Whatever the actual cause, by 1613 Thomas Milles, one of Jaggard's clients, noted that the printer had lost his eyesight completely. Jaggard's son Isaac was admitted that same year to the Stationers' Company, which oversaw printing, and assisted his father in future business ventures.

One of the most important of these ventures involved an extensively illustrated anatomy book by the London doctor Helkiah Crooke (a regrettable name for any doctor). On the market by 1615, *Mikrokosmographia: A Description of the Body of Man* immediately landed both Crooke and Jaggard in hot water for pornographic indecency—not so much for the "body of man" as the "body of wo-man."

Part of the publishing process in 1614 included submitting the text to the Bishop of London for censorship and approval. When Dr. Crooke did this, the Anglican bishop took particular offense at illustrations contained in "Of the naturall parts belonging to Generation," which is doctor-speak for "This is the section where we talk about penises and vaginas, so let's be adults please." Only, surprise, surprise, the dickpics wouldn't be the issue. Plate VI contains the drawing of a dissected female torso and abdomen, including the labeled reproductive system. Perhaps most scandalously in this medical encyclopedia is the "vaginal cleft in full anatomical detail."

The Bishop of London immediately took the book to the College of Physicians. The president of the college looked it over and issued his own verdict: indecency. But not for vaginal clefts. The *Mikrokosmographia* was the first comprehensive anatomy volume to be published in English rather than Latin. The president of the College of Physicians didn't like this one bit. If anyone could just crack a textbook and read the words, who knew what mayhem might ensue?

Even though England had won the battle for an English Bible, that was the Word of God, not pictures of naked ladyfolk. A responsible pedagogical illustration meant for educated men would transform into vulgar pornography if viewed by those who hadn't been properly trained. Only educated men could get away with looking at nudity and remain free from moral corruption. A book containing nude pictures *and* written in the common English tongue might fall into the hands of people who were never meant to read about vaginas—women, for example. If the *Mikrokosmographia* were published without "the naturall parts belonging to Generation" being excised, the president of the College of Physicians threatened to "burn it wherever he found it."

After attempts to intimidate Dr. Cooke failed, the college went after William Jaggard, summoning him to appear before an assembled court. But Jaggard was blind, and presumably didn't feel like it, so he sent his wife instead. The college yelled at her for a while, then sent her back to her husband, confident that Jaggard would change his mind about the printing. He did not. *Mikrokosmographia* was published in the spring of 1615 with no changes made whatsoever. As far as we know, the Bishop of London took no further action, the president of the College of Physicians did not become a serial book arsonist, and the city didn't collapse because women could read anatomically accurate descriptions of va-jay-jays.

Writers and scholars have at various times attempted to cast William Jaggard as a villain in the history of print. Charges of false advertising and forgery have followed him for hundreds of years. But there's another picture of Jaggard that has slowly emerged from the historical record: one in which we see a defiant old printer who stuck to his guns when it came to censorship and the rights of printers and publishers. Did he imply that Shakespeare was the author of the collected poems in *The Passionate Pilgrim*? Yes. Did he bite his thumb at the Bishop of London and the College of Physicians when they were squeamish about feminine clefts? Also yes. Did he go blind from an STD? Irrelevant. Did he fake the title pages in a collection of plays prominently featuring Shakespeare? Okay, this one deserves a little more explaining.

Sometime in the early months of 1619, rumors likely started circulating that Jaggard, along with a prominent stationer, Thomas Pavier, was printing a collection of plays primarily written by Shakespeare. As far as we know, the two made sure to secure rights for the individual plays from the printers who owned them. No one would have cared outside of two very prominent members of the theatrical community: John Heminges and Henry Condell, both of the King's Men players.

Heminges was the oldest member of the King's Men, and a close acting partner of Shakespeare's since the formation of the original Chamberlain's Men in 1594. Besides his work with Shakespeare's plays, Heminges is best known as the CFO of the acting troupe. He was one of the eight primary shareholders of the company, and quickly became their head accountant. By the time of his death, he personally held a quarter of the shares in both the Globe and the Blackfriars, the two theaters owned by the King's Men. And at a yearly salary of around two hundred pounds (the cost of attending university, for example, was about thirty pounds a year), Heminges was described in 1619 as a man of "great living wealth and power."

Few details are known about Heminges outside his interactions with the theater. Even less is known about his close financial partner,

Henry Condell. Condell first shows up in the historical record as an actor in 1599, and four years later we find his name prominently displayed on an official list of King James I's royal theatrical servants. He almost surely was among the twelve King's Men who in 1604 were invited to travel to Somerset House as the king's personal eye candy.

Although actors have not always sported the best reputations in society, in May 1603 (the year of the "bad" *Hamlet*), Shakespeare, Heminges, and Condell, among other King's Men players, were invited into the service of His Majesty's groom of the chamber. These lower-level courtiers were responsible for tasks such as handing royal clothing to a "squire of the body," who would then dress the king.

The title was largely ceremonial, but some of the King's Men did serve as glorified coatracks on at least one occasion when Spanish delegates arrived in London to negotiate peace with England in 1604. The King's Men were instructed to stand around in their new scarlet cloaks, doublets, and breeches for eighteen days and look pretty. They weren't to perform anything; they were just to demonstrate to the Spaniards that King James I could take very talented people and make them stand in one spot for two and a half weeks. In a receipt dated August 1604, the king commended his players for a job well done.

Heminges and Condell's close relationship with the Lord Chamberlain (who oversaw the groom of the chambers, and under whom the King's Men served as royal actors) proved highly advantageous when rumors started circulating of William Jaggard and Thomas Pavier's forthcoming Shakespeare collection. The Lord Chamberlain issued an edict in May 1619, suddenly making it illegal for anyone to publish a play of Shakespeare's without first securing the permission of the King's Men players. Despite holding "rights to copy" for many of the plays, it appeared that Jaggard and Pavier's project was shut down.

Then copies of the Shakespeare plays slotted for the collection began surfacing—as individually dated quartos. These quartos contained printing dates ranging from 1600 (twenty years earlier) to 1608 to 1619, just before the Lord Chamberlain's edict. The words "Printed for T.P." (Thomas Pavier) are found on six of the ten plays, and for that reason, they have been branded the Pavier Quartos. Except for two plays attributed to the work of "J. Roberts" in 1600, there are no immediately obvious indications as to who printed them.

For hundreds of years, it was believed that, after Heminges and Condell used their Crown connections to make printing Shakespeare illegal, Pavier abandoned the idea of a collected works and just started dumping his inventory. It was assumed that he had previously purchased surplus copies of Shakespeare play texts (printed in 1600 and 1608), which included such well-known plays as *The Merchant of Venice*, *A Midsummer Night's Dream*, and *King Lear.* His personal collection also included the plays *Sir John Oldcastle* and *A Yorkshire Tragedy.*

(Side note: if you took a Shakespeare class in college and don't recognize those last two titles, don't worry, you weren't so hungover that you forgot two whole Shakespeare plays. While *Sir John Oldcastle* and *A Yorkshire Tragedy* were at times thought to be the work of Shakespeare, they have since been definitively rejected from the canon as apocryphal.)

Presumably, Pavier halted the printing endeavor with Jaggard, sold off the last of his Shakespeare surplus, and then sat in a room with a slow-burning fire and a glass of sherry throwing darts at portraits of Heminges and Condell. But in 1908, W. W. Greg, one of the most influential Shakespeare scholars of the twentieth century, published *On Certain False Dates in Shakespearean Quartos*, and smashed the surplus theory to pieces.

Greg and other bibliographers demonstrated that, despite the claims on the title pages, all ten plays were printed together in 1619 in the same printing shop. Forensic evidence showed that paper from the same job stock was used in plays that were supposed to have been printed twenty years apart; that the unique watermarks created by the original papermakers didn't match the proposed dates (*Henry V*, for example, was supposed to have been printed in 1608, but "appeared to have a watermark dated 1617 or 1619"); and the personal stamps (or "devices") used by printers such as "J. Roberts" had clearly been stamped much more recently than claimed. (Even with the naked eye, it's easy to see that the stamp used in one of these quartos allegedly printed in 1600 is actually older and more weathered than the same stamp from 1605.)

At the dawn of the twentieth century, almost three hundred years after the Pavier Quartos were created, judgment was finally passed: forgery. Perhaps most damning is the sequencing of the quarto pages. Because they were intended to be part of a broader collection, the first three plays (the two parts of *The Whole Contention* [2 and 3 *Henry VI*] and *Pericles*) were printed together, and therefore carried continuous signatures. "Signatures" are small letters printed at the bottom of the pages as a guide for putting sheets in the right order when binding.

After *Pericles*, the presses abruptly halted. When they started back up again, the last seven plays were printed with their own individual signature sequences. This would very strongly suggest that something significant happened in early 1619 between the printing of plays one through three and plays four through ten. Something like a plague, or a fire, or a meddling Lord Chamberlain.

Today there's little doubt that William Jaggard was the silent printer behind the unlawful 1619 publications. Some scholars now suggest we should be calling this group the Jaggard Quartos instead of the Pavier Quartos. Not only did Jaggard continue to pursue the Shakespeare project, but he altered the title pages to make it appear that the plays had been printed before the Lord Chamberlain's ban.

So we find Jaggard, now blind and four years from the icy grip of death, biting his thumb at the authorities once again. What would have motivated the old printer to do this? Money? Sure, that's one explanation. Ideally, one makes money by printing things (fingers crossed). Though it's unlikely Jaggard thought he'd be able to recoup his printing costs on a scrabble of play texts before Death came a-knocking for him. Jaggard and Pavier have been branded literary pirates by early twentieth-century scholars (most notoriously by Greg's bibliographic buddy, A. W. Pollard), but recent studies have suggested a much different fight taking place.

It would be best to start with the evidence brought against William Jaggard. Did he forge the title pages? Yes. We absolutely know that. How? Because he *signed* the forgeries. Ladies and gentlemen, the prosecution rests because the defendant is clearly an idiot.

Or is he? When printing a book in this period, it was customary to include a unique ornament, or "device," on the title page. Below this stamp, a printer would record his name, the place of publication, and the date. The forged Pavier Quartos have the stamp, but below that, just "London" and a false date. Here's the thing, though, that stamp (McKerrow device #283, a rectangular woodcut with flowers in the center encircled by the Welsh motto "Heb Ddieu, Heb Ddim," or "Without God, without all") was Jaggard's personal device. He'd been using it since 1610, and most recently in an official catalogue of upcoming publications printed nine months before the forged quartos. In the catalogue, the stamp is included on the title page with "London / printed by W. Jaggard / 1618" beneath it.

Any printer, publisher, bookseller, or bookbinder in London would have looked at that specific stamp and known who printed the item. Was Jaggard really that stupid, to use his own personal seal on a work of forgery? Was his syphilitic mind finally just giving out on him? Possibly, but some scholars have suggested that Jaggard's stamp is an ornately designed *Go hang yourselves, you malt-horse drudges* from a printing curmudgeon who resented the King's Men and the intrusive Lord Chamberlain.

This royal interference didn't upset only Jaggard, but also the established practices and guidelines of the Stationers' Company, the organization that formed the spine of the entire Elizabethan book trade. The Stationers' Company was originally a guild of copyists, bookbinders, and booksellers that predated Gutenberg's press by about forty years. By 1557 it had secured a royal charter and become the trade company that oversaw all printing in London.

Copyright as we understand it didn't really exist in Elizabethan England, but probably the closest thing to it was the Stationers' Register, which kept track of who had purchased rights to the publication of specific works. The Stationers were given power to run their organization without significant royal oversight in exchange for policing their own members, and their charter required them to seize or restrict publications that weren't officially licensed. So, for centuries, the Stationers were the gatekeepers, preventing the London book industry from collapsing in on itself.

The only major exception to the Stationers' autonomy was the royal prerogative to grant patents, which allowed certain printers (or, in even rarer cases, the authors) themselves the exclusive "privilege" to print popular works such as the Bible or annual almanacs. Printers would insert the words *cum privilego* on their title pages to announce that they held these royal patents, which voided anyone else's previous rights to copy. This species of biased royal meddling usually doesn't sit well with normal folks. It makes them do strange things, such as dress up as Native Americans and dump perfectly good tea into harbors.

The Lord Chamberlain's edict of May 1619 read, "[N]o plays that his Majesties players do play shall be printed without the consent of some of them." To printers such as William Jaggard, these kinds of royal favors were nothing less than attacks on the very foundation of intellectual property as it was practiced in the seventeenth century.

Playwrights such as Shakespeare usually sold their play texts to an acting company, and as far as they were concerned, that was the end of it. When a person purchased and then registered a play with the Stationers' Company, he or she held the rights to print and distribute it. The play could also be inherited or traded or sold to other companies or private citizens. This was the case for Jaggard and Pavier, who paid for, or otherwise gained the rights to, six of the eleven plays in the proposed "collection" (2 and 3 *Henry VI*, *A Yorkshire Tragedy*, *Henry V*, *Sir John Oldcastle*, and Heywood's *Woman Killed*, found in some bound copies of this group). Of the remaining five, Jaggard and Pavier might have had loose claims to three of them (*Pericles*, *The Merchant of Venice*, and *A Midsummer Night's Dream*), and the rights to the last two (*The Merry Wives of Windsor* and *King Lear*) were obtained for publication from their legal owners.

While the Stationers' Company had no recourse against a royal edict that instantly made worthless any titles that Pavier had previously purchased, it *could* (and apparently did) look the other way while Jaggard took the printing underground and slapped a big *Suck it, you clotpoles!* on the title page. Heminges and Condell, who were not stationers, would probably have missed the insult. Anyone who understood the world of print would not.

Heminges and Condell may have had another motivation for tattling to the Lord Chamberlain in 1619. Either before Jaggard began printing the Pavier Quartos, or at least very soon thereafter, the leading duo of the King's Men started planning the release of their own collection of Shakespeare's plays.

SHAKESPEARE HAD died a few years before, and Heminges and Condell were committed to gathering up and succoring his plays as if they were his own dear "orphans." A moment's reflection would reveal what a deadbeat this would make Shakespeare, since he sold his kids for eight pounds apiece and then walked out on them forever. Nevertheless, Heminges and Condell stated they just wanted "to keep the memory of so worthy a friend and fellow alive." They also wanted to make sure you purchased a copy of their book: "The fate of all books depends upon your capacities," they would write in a special note to "the great variety of readers": "and not of your heads alone, but of your purses . . . stand for your privileges we know: to read, and censure. Do so, but buy it first. That doth best commend a book." They cared about preserving the reputation of their dear, departed friend, yes, but making money off the process was also their solemn duty to bear.

Much could be said of Heminges and Condell's business acumen. They were competent accountants, each owned a quarter of the King's Men theaters, and they had grown considerably wealthy. Heminges's own daughter, Thomasine, unsuccessfully sued him after he repossessed her dead husband's shareholdings, and Condell reportedly owned a country home. (Owning a summer home is considered ritzy no matter what era you live in.) More than anything else, though, their contributions to the Shakespeare canon have become their true legacy, and those contributions cannot be understated, no matter what copyright traditions they may have trampled along the way. We take jabs here and there, but ultimately we owe Heminges and Condell an unpayable debt of gratitude.

Around February 1622, printing began on *Mr. William Shakespeares Comedies, Histories, and Tragedies*, remembered by the world today simply as the First Folio. This landmark publication has been called, in classic canonical book-speech, "incomparably the most important work in the English language." Of the thirty-eight (or thirty-nine) plays known to be authored by Shakespeare, eighteen of them (almost half) appear here for the first time, meaning they still exist only because of this collection. Without the efforts of Heminges and Condell, there would be no *Macbeth*, no *Twelfth Night*, no *Julius Caesar.*

The printing of thirty-six full-length plays was no small undertaking, especially because the 866-page tome was going to be printed in the much larger, taller folio format (around $8^{1/2}$ x $13^{3/8}$ inches), rather than the usual quarto size (around $6^{3/4}$ x $8^{1/2}$ inches). For years, Shakespeare's plays had been printed not as books per se, but as little quarto pamphlets, roughly stitched up the side, not meant for the lasting reverence bestowed on stately bound books. If you look closely at many quartos that have survived today, you can still see the delicate stab holes made by the needles.

Not just any printer would have been able to pull off a collection of this magnitude. Heminges and Condell needed someone who had connections with copyright holders. They needed someone who was willing to blaze new trails in printing, and who had the economic security to follow through. They needed someone who wasn't afraid to stand up to mockery and intimidation. They needed William Jaggard.

Of the great ironies in Shakespeare studies, this collaboration between Jaggard and the King's Men to create one of the most important books in Western literature certainly tops the list. No one really knows how aware Heminges and Condell were of Jaggard's subversive printing practices, but it's likely the old bastard had flung around his share of colorful epithets regarding the two actors who, just a couple of years before, had so thoroughly assaulted the sovereignty of printing rights.

By the time the First Folio was started in 1622, Jaggard already had one foot in the grave, and perhaps that was for the best. He certainly would have been aware that all the plays from Pavier's aborted quartos made it into the King's Men's collection (minus the Heywood, the two apocryphal texts, and *Pericles*, which was only partially written by Shakespeare). Adding insult to injury, the preface written by Heminges and Condell contained this bold statement: "As where (before) you were abused with diverse stolen, and surreptitious copies, maimed, and deformed by the frauds and stealths of injurious imposters . . . [they] are now offered to your view cured, and perfect of their limbs."

Did Heminges and Condell know that Jaggard was one of those frauds and stealths who engaged in the amputation of their friend's limbs? Did they know he would have scoffed at the idea of copyright nullifiers accusing *him* of diverse stolen and surreptitious activity? Probably not. Heminges and Condell would have gained little from pissing off their printer. But those statements wouldn't have escaped Jaggard's attention. And who knows, perhaps he died in the fall of 1623, right before the release of the First Folio, out of pure spite. Or maybe it was the STDs.

Jaggard's son Isaac had been taking on increasing responsibilities at the printing shop since his father's eyesight failed ten years before. By 1620 he was likely running the everyday operations. On November 4, 1623, a few weeks after his father's death, Isaac had taken over the shop and received the title of Printer to the City of London in his father's place. He was more than capable of carrying on his father's legacy. Even so, the publication of Shakespeare's First Folio was a risky endeavor.

A visionary man named Edward Blount can be credited as one of the most important investors to help make Shakespeare's collection a reality. A printer himself, Blount believed in great literature and had a reputation as a "literary arbiter of taste." The man had the unique pleasure of publishing works by some of the most accomplished writers of his time—and of all time: Marlowe, Montaigne,

Cervantes. There's no doubt he was a deep admirer of literature or that he derived immense satisfaction from bringing great writers to print. That's a good thing, because it was entirely possible that this satisfaction might have been the only compensation he received. According to Shakespeare scholar David Kastan, "The commercial context of the folio must not be forgotten. Today it seems obvious to us that the volume was the necessary and appropriate memorial to England's greatest playwright, but at the time all that was clear to Blount and his partners was that they had undertaken an expensive publishing project with no certainty of recovering their considerable investment."

Yet what could be so risky about an anthology from the most popular writer in England? Well, let's start with the physical book itself. In particular, the decision to print it in folio format. Plays, ballads, and other silly works of pop culture were typically printed in smaller, cheaper quarto or octavo sizes. They were meant to be read and then misplaced, or dropped out a window, or thrown at your servants. Folio, on the other hand, was the most dignified format for Elizabethan publishing. Theological commentaries were printed as folios. Historical compilations and legal works were formatted as folios. Writers were admitted into the folio club, but only if they were very good or very dead: Homer, Aristotle, Pliny the Elder.

When Heminges, Condell, Jaggard, and Blount decided they needed to print Shakespeare's plays in folio format, they were sending an unmistakable message: Shakespeare, though dead just seven years, had already become a classic. The Bard could stand shoulder to shoulder, bookshelf to bookshelf, with the likes of Plato, Ovid, and Vergil. "He is not of an age, but for all time!" wrote fellow playwright Ben Jonson. Which is nice, but if you dared publish an author in folio format who was not deemed a giant among men, people would mock the hell out of you. Ben Jonson would have known; he tried it.

In 1616, the year Shakespeare died, Jonson attempted to raise dramatic arts to the heights of fine literature when he published his *Works*, a collection of his own poems, plays, and entertainments, in folio format. Critics condemned him for his arrogance. "Pray tell me, Ben, where doth the mystery lurk, / What others call a play you call a work" (which was presumably followed by a seventeenth-century mic drop).

If readers accepted the idea that a pop culture playwright was worthy of the highest honors in the world of print, then the folio might succeed. If not, people would look at the folio in the same way a modern observer might view a gold-embossed, leather-bound edition of a romance novel such as *Pirate's Ransom*, *A Recipe for Temptation*, or *Master of Desire* (all actual, and delightfully awful, book titles).

Even if people could be sold on the idea that Shakespeare should be regarded as one of the greatest English writers, the price of the folio itself still posed a significant hurdle. Printing was not cheap. For a nice copy of the folio, bound with calf boards, you'd be looking at about one pound retail. You could get the volume as unbound sheets, but that would only knock around five shillings off the sticker price.

One pound was a substantial amount of money in 1623, equal to about a month's supply of bread. It should be noted that if you were patient, Shakespeare's First Folio would eventually have paid out. In 1623, a First Folio was worth the equivalent of 44 loaves of bread. By 1923, that same folio was worth 96,000 loaves of bread (feeding a family for roughly 182 years).

The First Folio was completed in November, less than two years after printing began. The first recorded sale was to a Sir Edward Dering, on December 5, 1623. Just looking at the numbers, it seems that the publishers' gamble paid off, eventually. It wasn't *Harry Potter*, but Shakespeare's First Folio sold well enough to warrant a second edition, in 1632. "It broke no records, but selling out inside nine years was a respectable performance for a fairly expensive folio."

While the First Folio was commercially successful, we would be remiss if we didn't at least mention its curse. It has not escaped the gaze of history that most of those who participated in the creation of Shakespeare's Folio were dead within four years. William Jaggard died during the printing. Thomas Pavier died two years later. Both Isaac Jaggard and Henry Condell died two years after that, with John Heminges following in three years. Edward Blount held out the longest, finally succumbing in 1632, nine years after the First Folio was completed. Is this mere coincidence? Are we really to believe that six men of varying ages could just die of natural causes in early modern England? The answer is yes. Absolutely yes. This was a time before penicillin, or washing your hands, or English anatomy books that described private parts. It was unfortunate that none of the publishers lived to see the Shakespeare Folio sell out its first run, but hey, if you want to live past forty, don't be born in the sixteenth century.

Levity aside, there actually is a curse associated with Shakespeare. The epitaph on his grave reads: "Blest be the man that spares these stones, And curst be he that moves my bones." Evidently Shakespeare was not only a playwright, but also a pirate who cursed his gravestone. When the Holy Trinity Church in Stratford-upon-Avon was renovated in 2008, developers actively worked around the gravesite so as not to disturb the Bard's bones and bring upon themselves a swashbuckler's fate. In a classic Shakespearean twist, however, some of the Bard's super-fans decided to ignore the curse. The archaeologist Kevin Colls recently discovered evidence that Shakespeare's skull was likely stolen from its grave sometime in the nineteenth century. *Hey, there's nothing wrong with lovingly carrying around another human being's skull,* says Hamlet.

Based entirely on what we have left to us, it appears Shakespeare cared more about his skeleton than his *Hamlet.* He feared people touching his bones more than touching his plays. He stressed out over a section of church floor more than the Kingdom of Denmark. Shakespeare's final, paranoid words were etched in stone, but his plays would live on only by the grace of other people working to preserve them.

"Alas, poor Yorick. I knew him," Hamlet famously bemoans as he lifts Yorick's skull and presumably curses himself, "Where be your gibes now, your gambols, your songs, your flashes of merriment that were wont to set the table on a rore?" The answer: as far as we know, they are nowhere. Yorick is only briefly introduced in act 5, scene I, so his jibes and gambols and songs and merriment do not actually exist. Shakespeare never wrote them. And if people such as Heminges, Condell, Jaggard, and Blount hadn't intervened on their behalf, the actions of Macbeth or Petruchio or Caliban would likewise not exist, either.

"If Shakespeare cannot with any precision be called the creator of the [Folio] that bears his name, that [Folio] might be said to be the creator of Shakespeare," David Kastan has concluded. Whatever the motivations and the politics behind it, the printing of the First Folio in 1623 would eventually transform Shakespeare into a household name for centuries to come. Ben Jonson may have failed to bring literary respect to play texts, but where he failed, Shakespeare succeeded brilliantly, thanks in no small part to the printing press.

It may be impossible to say that we really know Shakespeare. What we think we know of him has been shaped by his admirers, his editors, his critics, and the printers who memorialized him. At one end of that spectrum stand Heminges and Condell, who claimed in the preface of the First Folio, "His mind and hand went together: And what he thought, he uttered with that easiness, that we have scarce received from him a blot in his papers." In their caricature of Shakespeare, he made no mistakes at all. From his mind, to his pen, to the annals of time went his plays whole and unblemished. If that were true of Shakespeare, he would be the only writer in the history of the human race to which that statement could apply. Even God handed Moses a rewrite on Sinai. But aren't Shakespeare's gifts all the more marvelous *because* we know that he was a flesh-and-blood, imperfect person, just like us?

Whatever his "true writings" may be, there is no doubt that Shakespeare stands as a giant of Western literature. With the publication of his complete plays in the First Folio, he was clearly accepted by his English contemporaries as a writer of extraordinary skill. For the four hundred years since, printing presses have influenced and carried his words to every corner of the earth. At times the picture that emerges of Shakespeare is one that "pusles the braine, and doth confound the sence." But, in the end, we would rather "beare those evilles we have, / Than flie to others that we know not of" (Q1, 1603, "bad" *Hamlet*, E1^{r}).

BENJAMIN FRANKLIN MAKES IT RAIN

WHEN BENJAMIN FRANKLIN WAS TWENTY-THREE years old, arrangements were made for him to settle down and find a wife. Living on his own in Philadelphia, young Benny had it all. He owned his own home, a small printing business, and just a few years before, he'd purchased a "genteel new suit," a fancy watch, and walked around with "near five pounds sterling" in his pockets. But does a home and a business and nice clothes and cash in your pockets mean that you've succeeded in Colonial America? It seems to be a pretty good yardstick for success in Philadelphia circa 2017. In 1729, however, Franklin needed a little more.

A wife was a necessity for a printer—not for the usual reasons of companionship, reproduction, and an excuse never to go on awkward first dates again. In eighteenth-century America, the profession of printing was a game of survival, and printers lunged at any economic advantage. One major advantage was a wife who kept the books, folded and quired printed sheets, ran the general store, and even pulled the press or composed lines of type. Franklin needed *that* kind of a wife.

A tenant of Franklin's, one Mrs. Godfrey, took it upon herself to find Benjamin a match, and she seems to have succeeded brilliantly. By all accounts, the prospective bride was intelligent, attractive, and capable. You might even say that young Franklin was falling in love. But love doesn't pay the bills, my friend, and Benjamin—he had bills.

As part of the negotiations, Franklin asked for a one-hundred-pound dowry to pay off his business debts. One hundred pounds was a bit over a year's salary when he was working as a manager at a previous print shop. So take your annual salary, beef it up with a little overtime, and that's the bill you hand your in-laws for taking their daughter off their hands. The young woman's parents came back to Franklin saying "they had no such sum to spare." Franklin suggested they might "mortgage their house." After a few days, they countered with "[we do] not approve of the match," which is the eighteenth-century equivalent of *aw, hell no.*

At first it might appear that Franklin had overplayed his hand, but in the list of objections to his pursuit of their daughter, the hefty dowry was not among her parents' grievances. Mrs. Godfrey later explained that, after talking to a local printer named Andrew Bradford, the young woman's parents were "informed the printing business was not a profitable one." This Mr. Bradford helped them see that expensive materials such as metal type broke down often in the printing business and had to be replaced from across the Atlantic, in England. Two previous printers in Philadelphia had been forced to close their doors, and Mr. Bradford was sure that Mr. Franklin would "soon follow them."

After making a few more attempts to meet eligible women and getting nowhere, Franklin had to admit two things: first, "the business of a printer [is] generally thought a poor one," and second, "I was not to expect money with a wife, unless with such a one as I should not otherwise think agreeable." This is the Colonial way of saying, *Printing is a dead-end job and the only marriage proposal I'm likely to secure is one founded on cooperative desperation.*

In the meantime, Franklin nursed his sorrows in the bosoms of paid escorts. "I hurried frequently into intrigues with low women . . . which were attended with some expense," and a "continual risk to my health . . . though by great good luck I escaped it." It's true that no matter how bad life gets you down, everything's worse with chlamydia.

Benjamin Franklin's conclusions in 1729 were probably a bit on the pessimistic side. Within a year or so of his failed engagement, he reunited with Deborah Read, a former sweetheart. Deborah was "good and faithful," but more importantly, she worked hard and efficiently alongside Franklin in his printing endeavors over the next thirty-four years (and at less risk of him contracting chlamydia. From her, anyway).

Though he couldn't have known it at twenty-three, young Benjamin stood on the precipice of something truly great. Over the next twenty years he would build a printing empire that extended all over the colonies. His actions would help transform the printing press on this continent into a juggernaut of political discourse and social reform. His almanacs would become a foundational symbol of the American experience, and his newspapers a vehicle for fanning the flames of revolution. Also, they would make him a lot of money—like, an ungodly amount of money. The kind of money that lets you retire at age forty-two and spend the next several decades sipping champagne with Parisian debutantes.

The obstacles to Benjamin Franklin's success were many. To create his empire, he had to follow the money, pinpointing the most lucrative opportunities and devising plans to undermine his competitors. By far the biggest name on his list was Andrew "the In-Law-Whisperer" Bradford. Two men enter Philadelphia. One man leaves alive. (That is a bit dramatic. Both men eventually died in Philadelphia, but "two men enter, one man retires and moves on to a life of public service" didn't have quite the same ring to it.)

FRANKLIN'S INTRODUCTION to the world of printing began when he was twelve years old. Born to a Boston candle maker, young Benjamin soon found that spending all day cutting wicks and filling molds was not to his liking. Josiah Franklin, shocked that his preteen would rather be swimming at the beach than chandling tallow or boiling soap, pressured the boy into signing a nine-year contract of indentured servitude to his older son, James.

Just the year before, James had returned from England with presses and types and everything a printer needed to start a newspaper. You might think that young Benjamin would be forever grateful to James for rescuing him from the monotony of candles and setting him up as an intern in the exciting and lucrative world of newspapering. In fact, Benjamin hated James when they were working together; he was a servant, not an intern; and apprenticing at a printing house was neither profitable nor remotely pleasant.

Not only were apprentices required to work long hours, six days a week with no pay, but they also had to swear an oath of personal virtue to their masters. This meant no gambling, no drinking, and absolutely no fornicating. Without the holy trinity of things young men do between the ages of fifteen and twenty-five, it's not surprising that the intern pool was somewhat shallow. Given the generally poor reputation surrounding the trade itself, one historian explains, "printers were often 'obliged to take the lowest people' for apprentices and journeymen, because no family 'of substance would ever put their sons to such an art.'"

(Side note: Indentured contracts were printed in pairs, on a single sheet of paper, then cut down the middle to produce two copies. The middle cut was serrated, thereby creating a unique edge of *indentations* used to verify that the contract of the apprentice matched the contract of his master. Hence, the "indentured" part of indentured servitude. See, history *is* fun. Not for them, but, you know, for us.)

The life of an indentured servant was not exactly glamorous. Nor did it command much respect. Here is an actual ad placed in a newspaper after one apprentice at the *Maryland Gazette* went missing: "[He is] very thick, stoops much, and has a down look; he is a little pock-pitted, has a scar on one of his temples, is much addicted to liquor, very talkative when drunk and remarkably stupid."

Bosses were always having to correct their apprentices. This was not a problem for James Franklin. "Though a brother, [James] considered himself as my master . . . I thought he demeaned me too much . . . [he] was passionate and had often beaten me, which I took extremely amiss."

Despite his brother's abuse, young Benjamin felt the pull of the ink and quill. He wanted to write for his brother's paper, *The New-England Courant*, but he knew that James and his fellow newspapermen would never take him seriously (to be fair to them, he *was* only sixteen). To subvert this, Franklin altered his handwriting and took on the guise of a widow pseudonymously named Silence Dogood. Every two weeks or so, Franklin/Dogood would slip an op-ed under the door of his brother's print shop. Totally unaware of the real author, James and his associates seemed only too delighted to publish Widow Dogood's letters. "They read it, commented on it in my hearing, and I had the exquisite pleasure, of finding it met with their approbation," Franklin would recall. Just a spoonful of tricking your boss as an insightful middle-aged woman with three kids makes the workplace beatings go down.

It wasn't just the printers who were fooled by Silence Dogood. She became well liked enough in Boston that several men reportedly wrote into the paper offering to, let's say, rectify her widow status. You can't blame them. Dogood was an insufferable tease. "[Widowhood] is a state I never much admired," wrote a sixteen-year-old boy somewhere by himself in a darkened room, "and I am apt to fancy that I could be easily persuaded to marry again, provided I was sure of a[n] . . . agreeable companion." That is how Benjamin Franklin, in 1722, catfished the entire city of Boston.

When James found out who was behind the Dogood letters, he was not pleased. It was one of the final breaking points between the brothers that led to Benjamin fleeing the city, presumably ahead of a mob of randy male readers of the *New-England Courant*.

At the height of Benjamin's Widow Dogood period, James got himself into a bit of hot water with the Massachusetts Assembly. In the June 1722 edition of his paper, he published an anonymous article criticizing how slowly the assemblymen had been dealing with piracy. The illicit hijacking of citizens' goods might not have moved the Assembly, but call them mean names in public and, by God, they will drop everything and make you pay. Failing to uncover the identity of the article's author quickly enough, the now thoroughly motivated Assembly arrested and jailed James as the proprietor of the offending newspaper.

In an age when partisan networks and celebrity pundits enjoy the freedom to ignore facts and flout journalistic integrity pretty much at their own discretion, we sometimes forget that, at its birth, the American press was a loyal and subservient arm of the government. Its purpose was solely to inform the public of their leaders' actions, not to challenge or question those leaders. In fact, doing so was often illegal. The idea of a free press wasn't even considered a virtue at the time. William Berkeley, a seventeenth-century colonial governor of Virginia, infamously remarked, "I thank God, there are no free schools nor printing, and I hope we shall never have these [for a] hundred years; for learning has brought disobedience, and heresy, and sects into the world, and printing has divulged them . . . God keep us from both!"

What happened to James Franklin in 1722 was the norm, not the exception, in the American colonies. Yet printing was legitimately needed for official documents, announcements, currency, dispersing laws, and the like. So how does a government (or a modern billionaire oligarchy, for that matter) keep its watchdog muzzled, leashed, and sedated?

Cash—and threats. But mostly cash.

Follow the money: the greatest leverage Colonial leaders had over their local presses was government printing contracts. As much as 90 percent of all printing in the colonies at this time fell under these types of contracts, and in many areas the printers worked exclusively at the invitation and with the blessing of their regional governments. This would be like a presidential administration today making it illegal for anyone besides a single hand-picked network to air press releases, addresses, and governmental news. Competition for contracts like these tends to carry the subtext *Please pick us. We swear we'll print whatever the hell you want.*

This afforded colonial governments immense control over any information released to the public. Publish an article critical of your Assembly or governor, and your contract gets revoked. If you didn't have a contract (see: James Franklin), then you could be arrested and prosecuted for "seditious libel."

Some of the first rumblings of what we now recognize as freedom of the press emerged from conflicts over these government contracts. Here's a typical scenario: The Pennsylvania Assembly awards its contract to Printer A. Printer B is like, *Shit, what am I supposed to print now?* The answer: anything you can to survive. In 1685, William Bradford (father of Andrew "the In-Law Whisperer") was chastised by Philadelphia magistrates for publishing an attack on the Quakers. Bradford shot back, saying you really couldn't blame him; there was so little to print in 1685 that he had to take work wherever he could find it. In another tussle with the government just a couple of years later, Bradford was told, "You know the laws, and they are against printing, and you shall print nothing without allowance."

When publishing his newspapers in Philadelphia decades later, Benjamin Franklin echoed this economic motivation for a free press. By operating a press that was open to everyone, you had substantially more clients and a more stable base of readers. High ideals and a sense of protected rights eventually merged with the pursuit of wealth, birthing our freedom of the press—which is a pretty good description of America's birth as well.

James Franklin was imprisoned for a month at the pleasure of the Massachusetts Assembly. When he was released, he was ordered to refrain from printing the *New-England Courant*. To get around this, he bequeathed the paper to Benjamin—or, at least that's what he wanted the Assembly to *think*. Nullifying the original indenture, James made a show of publicly releasing his younger brother from apprenticeship and turning the entire business over to him. Privately, however, Benjamin was only a front. In fact, James strong-armed his younger brother into signing a new and secret contract to operate the *Courant*, essentially indenturing Benjamin to himself as both the apprentice and the master of the same newspaper.

Franklin remembered this as a "very flimsy scheme," and it didn't take long before he'd had enough. Normally, indentured servants don't get to walk away from a job just because it doesn't pay money, or your coworkers beat you, or your boss calls you pock-pitted and remarkably stupid. But as the fake head of a newspaper, Benjamin had all the freedom in the world to kick open the front door and wave good-bye with his middle fingers raised high. Any attempts to enforce his indenture would have resulted in said indenture becoming public—something James could ill afford if he wanted to steer clear of Boston prison cells.

Besides, Benjamin had learned enough of the trade during his time with the *Courant* that he could easily find employment at another printing house. He didn't count on his brother going behind his back and bad-mouthing him to every printer in Boston. But, screw Boston; he could always go to New York. His father tried to prevent him from leaving, but Franklin arranged with the captain of a New York sloop to smuggle him out of the city. His cover story was that he'd "got a naughty girl with child, whose friends would compel me to marry her."

Finding no work for a printer in New York, Franklin eventually made his way to Philadelphia, where he became the employee of one Samuel Keimer, a failed printer from London. One of only two printers in the city (the other was Franklin's soon-to-be In-Law Whisperer), Keimer's shop was an unmitigated disaster. His press was "shattered," his font was "small [and] worn-out," and he seemed wholly ignorant of how to print something effectively on paper.

Still, at least he paid, and in silver sterling no less. Keimer had come to Philadelphia when the city was populous enough to sustain a second printer operating without government contracts. Because he had to find paying gigs outside of Lady Columbia's handouts, as a source of potential income, he looked to people who wanted to be in power but who *weren't*. By economic necessity, printers without government contracts opened their presses to partisan politics. The existence of more than one printer in a Colonial city was another important step in the birth of the American free press.

After setting Keimer's press in order, saving up a little of that sweet silver, and making friends with the governor of Pennsylvania, seventeen-year-old Benjamin boarded a vessel back to Boston. But this wasn't some tail-between-the-legs, middle-of-the-night homecoming. It was a midday ticker tape parade. He was carrying a letter from Governor William Keith detailing the governor's proposal to patronize the young Franklin's new printing shop, pending his father's approval. After seven months working hard in Philadelphia, Benny had stacks in his pocket and he was itchin' to make it rain.

"I went to see [James] at his printing-house: I was better dressed than ever while in his service, having a genteel new suit from head to foot, a watch, and my pockets lined with near five pounds sterling in silver. He received me not very frankly, looked me all over, and turned to his work again."

Walking up to James's employees, Benjamin continued his victory lap. "One of them asking what kind of money we had [in Philadelphia], I produced a handful of silver and spread it before them . . . Then I took an opportunity of letting them see my watch; and lastly, (my brother still grum and sullen) I gave them a piece of eight to drink, and took my leave." Spread your bling on the table, buy a round of beers, then pop and lock your way out the door.

Years later, James was still smarting from this scene. Asked by their mother why the two couldn't reconcile, Benjamin later wrote that "he said I had insulted him in such a manner before his people that he could never forget or forgive it." Because that's how eighteenth-century ballers do it, son.

Josiah Franklin rejected the Pennsylvania governor's proposal to set up Benjamin with his own shop. While generally impressed that his son had accomplished so much in the few months he'd been away, Josiah felt that the boy, at seventeen, was too young to be entrusted with his own business. (Embarrassing his older brother in his own shop probably didn't help, either.) Upon receiving Josiah's refusal back in Philadelphia, Governor Keith vehemently disagreed.

"There [is a] great difference in persons," the governor replied to Benjamin; "and discretion [does] not always accompany years, nor [is] youth always without it." And since Josiah was unwilling to aid in the establishment of his son's printing shop, "I will do it myself," said Keith.

Starting a business costs money, a lot of money. But the governor was enthusiastic, and eager to accommodate. After Franklin submitted a proposed inventory that amounted to about one hundred pounds sterling, Governor Keith suggested that the young printer travel personally to England to "choose the types, and see that everything was good of the kind." Franklin agreed, and Keith told him that he'd write up a letter of credit and send it with him on the next ship to London.

Now, discretion may not always accompany years, as Keith observed, but Franklin was either too young or too inexperienced to realize what most Americans instinctively know today: never trust a politician. Don't do it. Saying whatever his constituency wants to hear is pretty much what makes a politician a politician.

Upon his arrival in England, one of Franklin's friends let him in on Governor Keith's "character." "[He] told me there was not the least probability that [Keith] had written any letters for me; that no one, who knew him, had the smallest dependence on him; and he laughed at the notion of the governor's giving me a letter of credit, having, as he said, no credit to give."

Politicians! The embarrassment Franklin must have felt at trusting one would have been surpassed only by the galling realization that sometimes fathers are right and you are wrong.

Stranded in London without money or prospects, Franklin found work as an apprentice in the printing shop of one Samuel Palmer. Over the next year, he worked as a pressman and a compositor, rising through the ranks and demonstrating to his English counterparts that Americans are hardworking, technically capable, and insufferably dry.

"My companion at the press," wrote Franklin, "drank every day a pint before breakfast, a pint at breakfast with his bread and cheese, a pint between breakfast and dinner, a pint at dinner, a pint in the afternoon about six o'clock, and another when he had done his day's work."

That's, what, six pints of beer at work every day? Before, during, and after breakfast? It's like the alcoholic equivalent of hobbit meals. Pressmen in London were consuming enough alcohol on a daily basis to employ an "alehouse boy," whose only job was to ferry beer between the pubs and the printing houses.

Franklin, on the other hand, drank exactly no beers at work each day. Unused to such high volumes of alcohol flowing through the workplace, he looked upon the actions of his beer-guzzling coworkers as "detestable," and set out to change their ways. *Hey, bugger off, Benjamin Franklin!* says any American reader who wishes he could drink even *one* beer at work every day. But don't be too harsh on young Franklin, dear reader. If it weren't for him, you might still be British and saying things like "bugger off."

"On occasion I carried up and down stairs a large form of types in each hand, when others carried but one in both hands. They wondered to see from this and several instances that the Water-American, as they called me, was stronger than themselves, who drank strong beer!" Strong to drink is strong to labor: probably the singularly most successful workplace lie ever perpetuated upon English business owners.

Indeed, strength, for pressmen, was one of the main requirements of the job. They spent so much of their day doing the hard labor of pulling a press that contemporary cartoons referred to them as "horses." (On the other hand, compositors were later called "monkeys" because they sat hunched over and picked at small sorts of type from printing cases.)

Franklin "endeavored to convince [one fellow pressman] that the bodily strength afforded by beer could only be in proportion to the grain or flour of the barley dissolved in the water of which it was made; that there was more flour in a penny-worth of bread; and therefore, if he would eat that with a pint of water, it would give him more strength than a quart of beer." Nothing like a little science to put workplace alcoholism in its place. Or not. "He drank on," Franklin observed.

Young Benjamin probably *was* stronger than his English counterparts. Not only was he an energetic nineteen-year-old, but unlike most men in the 1700s, he swam, enjoyed fresh air and daily exercise, and ate a conscientiously healthy diet. He could haul twice the weight that other pressmen carried, and when he was promoted to compositor, he set his type with an "uncommon quickness," which one could at least partially attribute to the alcohol-free clarity of his mind.

But why set type so fast? Why be the office ass-kisser? Oh, right, for the money. When you're a compositor and you set type faster than anyone else, you get paid more. In most cases, compositors were paid by the number of lines of type they set. The faster you worked, the more was sent your way. When you were fast, "job" printing (essentially single-sheet compositions) and time-sensitive projects, which paid noticeably better, landed on your desk first. Add to this the copious amounts of disposable income you didn't waste at the pubs, and you can see the financial advantages to being the one sober printer in the office. Their paying a tab of four or five shillings out of their wages every Saturday night caused Franklin to remark of his coworkers, "And thus these poor devils keep themselves always under."

Still, Franklin paid for his responsible behavior in other ways. When you voluntarily withdraw from the office beer fund, for example, you can expect your fellow compositors to retaliate. In Franklin's day, compositors selected letters from cases that were divided into separate sections for each letter of type, known as a "sort." All the *a* sorts were in one section, the *b* sorts in another, and so on. The sorts for capital letters were further separated, at the topmost section of the case, thereby creating the terms *uppercase* and *lowercase*. Compositors knew their cases so well that they could pick sorts from each section without even looking, just as we type on a keyboard without staring at the keys. Mysteriously, Franklin's sorts kept getting mixed up, greatly slowing his work. Disgruntled pranksters found other little ways of tripping him up as well. They transposed his pages. They broke his equipment. Sometimes they blamed the slipups on a ghost that was haunting the printing house. Being the office teetotaler was starting to cost Franklin in both the goodwill of his colleagues and in actual money.

Realizing his mistake, Franklin tried to build relationships with his colleagues in other ways. In place of their prebreakfast and breakfast ales, he helped his coworkers substitute the following meal: "a large porringer of hot water-gruel, sprinkled with pepper, crumbed with bread"—all for the three half-pence price of a pint of beer. *Cast away your ale, brother printers, I present you this bowl of watered-down oat flour and soggy bread.* (You just can't go back far enough in time to find English food that isn't bloody awful.)

After a year in Palmer's shop, young Franklin finally made arrangements to return to Philadelphia. Once again taking up work with the incompetent printer Keimer, he soon realized he had outgrown his former employer. Why try to keep Keimer's rotten ship afloat when he could build a new vessel and steer his own course? All he needed was two hundred pounds, or about two and a half years' of his journeyman salary, to get things started.

Compared to the lightning-fast service and reasonable shipping and handling fees of the twenty-first century, Franklin's acquisition of a printing press seems positively Stone Age. For one, the colonies weren't even close to self-sufficient. If you wanted a printing press, you had no choice but to order one from abroad. You couldn't skip down to the local expo and select the latest two-pull printing press showcased by bikini-clad tavern women. American-made products were universally recognized as expensive to produce, of limited quantity, and vastly inferior to their foreign counterparts. So poorly was Colonial industry regarded that even the idea of an American Bible was suspect; everyone assumed they would be less accurate than European Bibles. This belief was held not just by readers and clergymen from around Europe, but by Americans. It wasn't until the middle of the American Revolution that a complete Bible in English would proudly be published by the patriot printer Robert Aitken. This 1781 edition issued in Philadelphia, known in the rare book world as the Aitken Bible, is coveted by collectors of Americana and Bibles alike, and can reach prices of $150,000 or more.

In addition to the presses themselves, paper and metal type also had to be ordered and shipped from England. Consider that, in the seventeen years Franklin personally ran his shops in Philadelphia, he purchased four thousand pounds of metal type. In any given shipment, thirty pounds were just replacement quotation marks. The first type foundry established on American soil didn't even open its doors until 1796, thirteen years after the end of the American Revolution.

The paper situation wasn't any more encouraging. At this time, the best-quality paper came from the British Isles. That wasn't so bad, as long as you didn't mind waiting enormous shipping times and potentially incurring water damage along the way. It was a common enough problem that in one contemporary advertisement, a binder in the colonies boasted of his ability "to bind books neatly and to take salt water out of books."

It wasn't very easy to acquire materials for binding books, either. Binding was an expensive addition to a product that was already expensive to produce. For this reason, books were much more commonly bound in sheepskin in the colonies, a notably cheaper option than the typical calfskin. Sheepskin comes with its own problems, however. The poorer-quality bindings don't age well. As a result, a rare book dealer today can look at a shelf of worn sheepskin bindings and make a pretty good guess that it's a collection of eighteenth-century American imprints.

Scarcity of materials was simply a way of life for Colonial printers. Broken type, poor-quality paper, and worn sheep bindings might have made American books look cheap and unimpressive next to their London counterparts, but that was the reality of printing in America. In some cases, these problems have even translated into high prices within the rare book market today. Many American imprints were issued "stitched"—that is, without any of that expensive binding, but simply with the sheets sewn together. Ironically, most eighteenth-century American pamphlets carry a premium for collectors if they remain "stitched as issued."

The two hundred pounds sterling that Benjamin Franklin needed to open his own shop was not an easy sum to come by, but with the help of friends and investors (and despite one failed engagement), he was able to scrape together enough cash to make it happen. Now he was free to pursue the printer's ultimate goal, which was documented as early as 1534: "nearly all master PRINTERS STRIVE first of all AFTER PROFIT." But even with a printing press in hand, Franklin couldn't just slap ink to type and make a living. Two printers stood firmly in his way: Andrew "the In-Law Whisperer" Bradford, who owned the government printing contracts; and Samuel Keimer, who swept up the leftover table scraps. Franklin had to break their stranglehold on Philadelphia, or go under.

The first step was to destroy his former employer. Three years before Franklin opened his own shop, Keimer contracted with the Philadelphia Quakers (or Friends, as they were known) to print *The History of the Quakers*. The tome was destined to be the largest and most expensive book printed in Philadelphia up to that point. At 722 pages, it was the size of a briefcase and may have contained more pages than all the books Franklin would later print over his twenty-year career. Keimer had been dragging his feet on this behemoth, so when Franklin opened his doors in 1728, he went behind Keimer's back and stole the contract. It paid off, too. Franklin printed the last quarter of the book in about two months. Considering that there were five hundred copies printed for this edition, it would have taken a total of 361,000 individual pulls of the printing press to obtain every sheet for *The History of the Quakers*. If Keimer had been able to accomplish this at Franklin's rate, he would have finished the Quaker history in nine months, not the three-plus years that had already stacked up.

The History of the Quakers was a brilliant success for Franklin. Not only was his work noticeably better than Keimer's (which is funny, because Franklin worked on the first three-quarters of this book while in Keimer's employ), but the speed with which it was printed afforded his business "character and credit." As James N. Green and Peter Stallybrass put it, "the ostentatious, almost superhuman speed with which [Franklin and his associates] finished the job was designed to show everyone how industrious they were and how slothful Keimer had been all along."

That was one nail in Keimer's coffin. The second came with Franklin's decision, in 1729, to begin printing a newspaper. Philadelphia already had a newspaper, Andrew Bradford's *American Weekly Mercury*. According to Franklin, it was "a paltry thing; wretchedly managed, and no way entertaining," but it was a newspaper nonetheless, with words and news and stuff. Why would anyone need more than one newspaper in a city? Back in 1720, when Franklin's older brother started the *New-England Courant*, concerned friends made the argument that the American colonies already had a newspaper, the *Boston News-Letter*. "One newspaper [was], in their judgment, enough for America."

Getting wind of Franklin's plans, Keimer retaliated against the younger printer's scheming and started his own newspaper first, the *Universal Instructor in all Arts and Sciences: and Pennsylvania Gazette*. As it turns out, Keimer was just as bad at running a newspaper as naming one. Wracking his brain for something that might give him an edge over Bradford's *Mercury*, he came up with the brilliant (not brilliant) idea to headline each issue with an entry from a 1728 encyclopedia. Front-page news of issue number one would have been the lexical entry for "A." Yes, just "A: "A vowel, and the first letter of the *English* alphabet. See LETTER, VOWEL, and ALPHABET; *where what relates to A, considered in each of those capacities, is delivered*."

Whew. People in the 1720s found that about as interesting as you just did. Franklin publicly observed that, at the rate Keimer was publishing his newspaper, "It will probably be fifty years before the whole can be gone through"—if that were something a person would even want to do, which it was not. In the fall of 1729, struggling to keep the paper afloat, Keimer gave in and sold it to Franklin for pennies on the dollar. And in case you're wondering . . . "Air." That's how far Keimer made it: Air.

Within a few months of selling out to Franklin (who mercifully shortened the paper's name to the *Pennsylvania Gazette*), Keimer sold his shop to his apprentice and moved to Barbados. His apprentice couldn't do any better and closed down within the year. That left just Andrew Bradford, the man who had very recently convinced Franklin's potential in-laws to abandon their matrimonial negotiations. Time for some payback.

DISMANTLING BRADFORD'S organization would not be easy. Franklin had to follow the money, and that trail led right to the government contracts, delivery routes, and paper mills owned by the Bradford family.

Good morning, Mr. Franklin.

Your mission, should you choose to accept it, is to undermine Andrew Bradford's printing syndicate. This cheap Colonial

piece of paper will self-destruct on its own in five seconds.
Good luck.

The first step in taking down Bradford was diverting that sweet, sweet government printing. Franklin's plan was simple: let his work speak for itself. In March of 1729, as the official printer to the Assembly of Pennsylvania, Bradford printed the speech of Governor Patrick Gordon. Maybe familiarity and a false sense of job security caused Bradford to cut corners. The text he printed was a mess, full of clumsily arranged, broken, and missing letters. One look at the printed speech, and the Assembly would have been compelled to conclude, *Clearly this man does not give a shit.* But Franklin did give a shit, and he demonstrated it by reprinting the governor's speech using aesthetically pleasing arrangements and fonts, and bold, clean, accurate letters. After Franklin circulated his copies for comparison (and with a bit of help from political allies), the Assembly voted to turn the government contracts over to him. Speeches, broadside proclamations, minutes of meetings, and official government news and documents all had to pass through his shop now.

Franklin's newspaper stood to be his biggest moneymaker. The *Gazette* was objectively a better paper than Bradford's *Mercury.* In addition to the stream of European headlines, Franklin reported news from neighboring colonies and strove for higher-quality offerings than his competitor. His newspaper introduced a unique blend of instructive and entertaining writing. "As I know the mob hate instruction, and the generality would never read beyond the first line of my lectures . . . If I can now and then . . . satirize a little . . . the expectation of meeting with such a gratification, will induce many to read me through, who would otherwise proceed immediately to the foreign news."

But quality does not by itself guarantee selling power. If consumers can't access your product, they aren't going to buy it. This is where Bradford's second stranglehold came into play: he was the postmaster of Philadelphia. The postmaster controlled the main distribution networks throughout the colonies. When Franklin began printing the *Philadelphia Gazette* in 1729, Bradford slapped a ban on any of his post riders carrying Franklin's newspaper.

Franklin's paper needed a steady stream of revenue from advertising. A lot of money was made using subscriber numbers to attract advertisers. One Colonial printer called advertisements "the life of a paper." Because Andrew Bradford "kept the post-office . . . his paper was thought a better distributer of advertisements than mine, and therefore had many more, which was a profitable thing to him and a disadvantage to me."

Advertisers want the biggest pool of potential consumers. Distribution determined the size of that pool. Without the postmastership, Franklin was severely handicapped. Not only was he missing out on advertising revenue, but he was having to dig into his own pockets to pay off Bradford's postmen—"what [newspapers, etc.] I did send was by bribing the riders, who took them privately."

Whatever finagling happened behind the scenes is lost to history, but all Franklin really needed to do to break Bradford's postal siege was wait. In 1737, Colonel Spotswood, acting Colonial postmaster general, called on Bradford for an accounting of his books. Franklin uses words such as *negligence*, *inexactitude*, and *lack of clearness and punctuality*, when describing the reasons for Bradford's sacking. "Inexactitude" sounds an awful lot like "embezzlementitude" to us. Whatever the details may have been, Bradford was fired, and the job was offered to none other than Benjamin "Whose Printing Business Ain't a Profitable One Now?" Franklin. The issue of the *Gazette* that appeared just after this promotion contained twice as many ads as the one preceding it. By the time Franklin left the *Gazette* in 1748, subscriptions had jumped from ninety to fifteen hundred, and paid out eight hundred pounds per year. Advertising accounted for an additional two hundred pounds a year (roughly the cost of opening a whole new print shop annually). Together, these constituted more than 50 percent of Franklin's income at his Philadelphia shop.

Next, the problem of paper. Ordering paper from England meant waiting out the enormous shipping times, and it carried the risk of incurring water damage. Purchasing from a local dealer would have been so much easier. The problem was, there weren't many paper mills in the colonies. In Philadelphia there was only one, established in 1690 through a joint venture between the Rittenhouse family and William Bradford, father of Andrew. The Bradfords used their influence over the Rittenhouses to cut off paper to anyone they felt posed an economic threat to their organization. Back in the mid-1720s, Andrew Bradford forced Nicholas Rittenhouse and his son-in-law John Gorgas to stop supplying Keimer with local paper when he was attempting to print *The History of the Quakers*. They targeted Franklin in a similar manner, "forcing Franklin to use imported paper in everything he printed." Considering the hazards of shipping, on top of poor exchange rates for colonists, imported paper could cut heavily into an American printer's profits.

In 1733, Franklin was so desperate for paper that he had to go halfsies with his Bradfordian nemesis on a psalmbook because, without Rittenhouse's support, he wouldn't have had enough paper to finish the printing. This appears to have been the final straw for Franklin. A few months later, he began to setup a new local paper mill.

By the end of his career, Franklin claimed to have established eighteen paper mills in the colonies. Between 1739 and 1747, his ledgers recorded sales of eighty-three tons of rags sent to seven papermakers throughout Philadelphia, earning him in excess of a thousand pounds sterling silver. Franklin became the Godfather of American paper. In keeping with this theme, a wise man once said, "Keep your friends close and your enemies closer." But Franklin's nemesis was not a wise man. By attempting to choke out his competitors through a backroom paper drought, Andrew Bradford inadvertently created a monster, a papermaking Corleone monster with English silver running through its veins. *It's just business, Mr. Bradford, nothing personal. Well, you did break up my engagement and try to push my press out of Philadelphia, so, yeah, I guess it's a* little *personal.*

Government printing, postal distribution, and paper—these three haymakers certainly put Bradford off his feet, but it was the smaller body blows that really made Franklin the most successful printer in the colonies. The *Gazette* was going strong, and *Poor Richard's Almanack* was introduced in 1732, selling thousands of copies a year for the next twenty-five years. Almanacs were immensely popular; print historian James Raven notes that in eighteenth-century England "more copies of almanacs were sold than all other types of publication put together." These pamphlets contained articles of varying utility, from calendars, to weather forecasts, catalogues of British kings, dates of fairs, descriptions of road systems, nuggets of wisdom to be cross-stitched onto throw pillows, financial interest tables, and chronologies of "things remarkable." They were also the source for most of your favorite Franklin quotes, such as "Keep your eyes wide open before marriage, half shut afterwards," "He that drinks fast, pays slow," and "Three may keep a secret, if two of them are dead."

Franklin didn't stop there. He knew how to "[contrive] a copperplate press," which helped him to win contracts for printing currency in Philadelphia, New Jersey, and Delaware. He made his own ink. He could "cast his own [metal] type when sorts were lacking, something no one else in America could do." He eventually became the largest wholesale and retail bookseller in the colonies, dominating the transatlantic book trade. In an environment of chronic scarcity, the ability to create, control, and divert the American printing supply line was practically the equivalent of being able to print money, which he was also legally allowed to do.

But in addition to the high-minded newspapers, government speeches, and English literature, Franklin's second-largest source of income was from small incidental contracts, or job printing. These were your blank forms, your handbills, bills of sales, apprentice indentures, powers of attorney, bail bonds; even labels for medicine bottles and "wrapping papers for soap and tobacco." In the history of print, job printing is the intelligent but ugly older sister. Considering its value, job printing deserves a lot more attention than it gets, but this kind of printing hasn't historically been as sexy to scholars and collectors as books.

Jobbing work comes in an enormous variety, from orders for four thousand salad oil advertisements to the first printing of the Declaration of Independence. They are usually printed cheaply and have a devastating rate of survival, which is all the more depressing when you consider what we can potentially learn from them. Many of these printed materials would have seemed utterly mundane to an eighteenth-century audience, but today they serve as valuable windows into the past, like printed receipts, lottery tickets, advertisements, and labels.

Yet, herein lies the problem. Most of this work seems so unremarkable or so fleeting that it doesn't always register on our radar as important. The Dunlap broadside, the first printed copy of the Declaration of Independence, is one of those extravagant exceptions that proves the rule. In 1989 the twenty-fifth known surviving copy of the Declaration, printed by John Dunlap on the night of July 4, 1776, was discovered within the frame of an old painting bought for four dollars at a flea market. The document sold at Sotheby's for $2.42 million in 1991; then it sold for $8.14 million when put up for auction again in 2000. On the other hand, most ephemera have typically sold for zero dollars at auction because such humble sheets aren't deemed important enough to buyers. There are some exceptions, such as the nineteenth-century advertisement for carpet that artist William Blake was commissioned to illustrate. (This is a real thing: one scholar calls it "surely the most delightful carpet advertisement ever created.") But would a collector normally be interested in an otherwise random advertisement for carpet from 1820? No, which is our loss. But as historians and collectors attach growing importance to this seemingly modest detritus of print, we're seeing prices change and dealers handle job printing more. This is an important and exciting trend.

During the early years of the printing press, job printing sustained many printers. Given the time, labor, and materials required for these smaller printing jobs, they tended to be proportionally more profitable than projects we assume to be a printer's bread and butter. This makes sense when you consider the financial risks of a larger project such as a book. With capital tied into long-term investments, printers "needed the relief of regular, quick turnover, relatively simple jobbing." If Franklin sniffed out a path to make money, he followed it. These "little jobs" meant ready cash, and they were so important to Franklin's business that he literally stopped the presses whenever an order came in.

Add to this the hefty revenue from paper and the general supplies sold out of his print shop (which could include not only stationery, but items such as medicine, chocolate, whalebone, pickled sturgeon, and Spanish snuff), and you can see that Franklin had cornered the market in almost every aspect of profitable Colonial printing. Soon Franklin was "indisputably the dominant printer in Philadelphia and the wealthiest in all the colonies."

In 1742, after duking it out with Franklin for almost fifteen years, Andrew Bradford's health took a turn for the worse. He died on November 24, at the age of fifty-six, leaving the remaining shards of his Philadelphia printing empire to be divvied up among his living heirs.

Franklin remained in the business for only another six years before he'd made enough money to retire extravagantly. They say the first million dollars is the hardest to make, and that saying is not a new one. Well, the "million" part is. "My business was now continually augmenting," Franklin wrote, "and my circumstances growing daily easier, my newspaper having become very profitable . . . I experienced too the truth of the observation 'that after getting the first hundred pound, it is more easy to get the second.'"

With that kind of disposable income, Franklin invested in other Colonial printers. "In 1733[,] I sent one of my journeymen to Charleston, South Carolina . . . I furnished him with a press and letters, on an agreement of partnership, by which I was to receive one-third of the profits of the business, paying one-third of the expense."

With similar partners and associates in more than two dozen locations, including New York, New Haven, Lancaster, Newport, Annapolis, Williamsburg, Boston, and the Caribbean island of Antigua, Franklin created an actual news network that stretched out over the American colonies and beyond. The result was, as Green and Stallybrass have explained, "these partners and protégés all imitated Franklin's business strategies, creating a sophisticated intercolonial communications network, one of the most dynamic in the world, with Franklin at the center."

Franklin's network was the eighteenth-century version of an information superhighway. It was ideally placed to collect and spread this information. Unlike London printers, who had the economic luxury of focusing only on printed products if they chose, Colonial scarcity required printers to diversify. This meant they were often involved in multiple businesses in the community, such as a coffeehouse and general store, or Franklin's post office. (Thus the trend of bookstores with cafés, bemoaned by some bibliophiles today, is actually part of a long-standing tradition!) Among all these social and business connections, the local print shop would often become a gathering place for that community. Their newspapers became, as one contemporary described it, "vehicles of discussion, in which the principles of government, the interests of nations, the spirit and tendency of public measures, and the public and private characters of individuals, are all arraigned, tried, and decided."

Rather than a collection of loose colonies, these networks facilitated the idea that Americans were becoming a unified group. Now when something happened in New York, it carried weight in South Carolina. When Bostonians protested unfair taxation by dumping tea into their harbor, and England responded by sending in the army, Americans knew that "what was happening in Massachusetts could happen to them." Printing networks created a feeling that the colonies were all in this together. They were the crackling nervous system of the emerging body politic.

In the May 9, 1754, edition of the *Pennsylvania Gazette,* Franklin included a woodcut illustration of a snake segmented into eight parts. Below was printed the phrase "Join, or Die." This iconic image originally had nothing to do with American Independence. It was a call for unity during the French and Indian War to aid England in its struggle against France. Eleven years later, that same cartoon would be repurposed by William Goddard in the *Constitutional Courant* "as a symbol of colonial opposition to the Stamp Act in particular and British authority in general." Goddard was one of the printers in Franklin's network, and this network was ready to take action to defend itself when threatened.

The 1765 Stamp Act, a tax on paper goods in the American colonies to help pay for the French and Indian War, features prominently in the history of utterly stupid economic decisions. In theory, it seemed practical. A stamp act had been enforced in England in 1712, and no one revolted. Yet, when it was America's turn, Parliament hadn't considered the problem of chronic scarcity of materials. Many in the trade were barely scraping by as it was. Paper was the most expensive aspect of printing, and could annually exceed the cost of opening up a print shop. Now the British had slapped a tax on the single biggest monetary risk of printers' precarious livelihood, and they were pissed off. By virtue of their career choice, they had a ready-made platform for venting their righteous indignation.

Newspapers across the colonies united in what can arguably be called the first American propaganda campaign. How did principles such as taxation without representation become so villainized in the public consciousness? Follow the money. In the middle of those debates were the printers, protesting the price-gouging Stamp Act, equating it to the ramblings of tyrannical English overlords.

In the pre-Revolutionary furor that swept the colonies, the number of publications doubled. Printers who stayed true to England to win conservative government contracts faced potential disaster when angry mobs came banging on their doors. In one memorable case, a mob stole large quantities of metal type from James Rivington's Loyalist paper, the *New-York Gazetteer*, and melted it down to form bullets. Rivington fled New York soon after his house was torched by another mob, and then returned in 1777 to run the Tory *Royal Gazette*, while acting (or continuing to act?) as an agent for George Washington's Culper Spy Ring.

Franklin himself was initially slow to endorse American independence over reconciliation with the British Empire. Living in London at the time as the unofficial ambassador of the American colonies, he was somewhat out of touch with the changing resolve of his countrymen. Notwithstanding, the appropriated "Join, or Die" woodcut came to be viewed in England as one of the most radical publications from America, and "Europeans often credited [Franklin] with starting the American Revolution." It was as if Franklin's massive printing network went all Skynet on the eighteenth century and became self-aware. Where Franklin preached patience and rapprochement, newspapers such as the *Courant* cried, *Viva la revolución!* Printing had stepped past its master, marching the American colonies toward liberty and independence.

Franklin did eventually come around. In an essay reprinted in Goddard's newspaper, he sided with the anger of his fellow Americans over issues such as taxation without representation and the unjust reparations for damage done to the English economy. He publicly compared the situation—and this is real—to a Frenchman who holds a fireplace poker over a flame and then asks if he can insert said poker into the asshole of an Englishman. When the Englishman refuses, the Frenchman becomes irate and demands "payment for the trouble and expense of heating the iron." According to Franklin, that's what England was doing: attempting to anally penetrate Americans against their will (there is a word for that), and when the Americans resisted, England demanded reimbursement for the cost of the attempted assault.

Hit first in the money bags, printers followed their ideals and stepped beyond their supposed role as "mere mechanics" to fan the flames of revolution. Franklin's printing network was one crucial stage upon which the struggle for American independence played out. From this dais, "life, liberty, and the pursuit of happiness" became more than words—they became a foundation. What was once a question of cash soon became a quest for rights self-evident.

7

ANGELIC VISIONS AND DEADLY TERRORS

IN EIGHTEENTH-CENTURY ENGLAND, AN ENGRAVER was "the photographer of his day," hired to create reproducible prints for the public. As these prints became more fashionable, an entire industry sprang up around translating an artist's canvas into metal and ink. This kind of work required tremendous skill. One English engraver, for example, gained notoriety for how realistically he could reproduce cat hair. On the edges of this industry stood William Blake: poet, artist, and professional etcher. He didn't want his creations outsourced to another engraver, no matter how faithfully the man might carve cat hair into metal plates. Blake wanted to paint, engrave, and print that cat hair—or, more specifically, that Tyger hair—himself.

Amid the dark fumes and hissing acid, surrounded with turpentine, aqua fortis, wax, charcoal, and copper, William Blake must have resembled a mad scientist when he achieved his breakthrough. Sometime in early 1787, after years of experimenting with nitric acid to burn images into copper plates, Blake discovered a novel way to illustrate his poems. He and his younger brother Robert had been trying to crack the riddle of "relief etching," a graphic technique that would allow illustrated books to be printed quicker and more cheaply than traditional methods. For years, Blake had searched for the answer, feeling he "must create a [new] system or be enslaved by another man's."

The "system" most likely used if a publisher wanted to add a high-quality illustration in 1787 was intaglio printing. This meant hiring a professional engraver to cut the desired image into a copper plate using metal tools, or to etch the design chemically, using a complex process requiring wax, needles, and nitric acid. Both these methods were time-consuming and expensive, and required an entirely different printing press from the rest of the book. Like supermodels, it took a small army of talented professionals to make them look that good.

But Blake wanted total control, from conception through to the printing. No other compositors, inkers and pullers, editors, or publishers—just one simple line running from the artist to his work to the public at large.

After years of dissatisfaction, William and Robert Blake sat up one night talking in their London home and stumbled upon an elegantly simple solution. Instead of covering a copper plate in wax, using a needle to scratch out the intended design and "biting" the lines with acid, what if they painted directly onto the copper plate itself and then bathed the plate in acid to burn away the surrounding copper? This would create a plate whose design stood up off the surface of the metal in relief, like a modern-day rubber stamp. This method could potentially become a much faster and less expensive way to print any illustration an artist could imagine.

Relief printing was in fact an older form of graphic illustration—older than the printing press itself. In the early 1400s, woodcuts were used to illustrate everything from saintly icons (for praying) to sinful cards (for playing). Yet this type of printing required cutting away the wooden sides of the line you wanted to print, and thus produced a somewhat unrefined look. One scholar compares woodcuts to homemade gin: "cheap, crude, and effective."

Blake's proposed relief etching would dramatically reduce the cost of creating illustrations while still allowing the artist complete control over the brushstrokes. "Even Milton and Shakespeare could not publish their own works," Blake lamented. All he and Robert needed to do was find a viscous ink that would dry upon application, adhere strongly to metal, and resist all forms of nitric acid. No problem! Well, as long as you ignore the fact that such an ink did not exist, no problem!

Despite being William's junior in the world of print, Robert Blake actually came to his older brother's rescue and taught him a secret recipe for creating just such an ink. It consisted of precise ratios of turpentine, asphaltum (black petroleum), and linseed oil. High-fives all around! The next morning, William's wife ran out to the market with their last shillings and purchased the required ingredients—and it worked, just as Robert said it would.

If the story of Blake's relief etching ended there, it might be remembered as a mildly interesting anecdote about two brothers who invented something cool in the history of print. But there is more information you should know: Robert Blake was dead at the time. He had died earlier in the year, sometime before February 11, 1787.

Absent that key detail—that Robert's body was cold and buried in a London cemetery when he taught his brother the secret of relief etching—changes the nature of the story a bit. As it should. William Blake was an incredible poet, artist, and printer, but he was also considered mad, or at least delusional, by those who knew him best. Chatting the night away with the disembodied spirits of Shakespeare, John Milton, William Wallace, the biblical prophet Isaiah—this was all totally and monotonously normal for Blake. After all, why create new printing techniques by trial and error when you can simply learn about them from dead people?

Any change, large or small, in the details of a story can translate into significant changes to its overall meaning. This applies to more than just history. It also applies to books. And it especially applies to the works of William Blake.

SONGS OF INNOCENCE

1789

The Author & Printer W Blake

This is the text of the title page in Blake's most famous printing endeavor. *Songs of Innocence* is a collection of poems (or songs) apparently intended for children. It includes such titles as "A Cradle Song," "Infant Joy," "The Lamb," and (because this was the eighteenth century and kids got bills to pay) "The Chimney Sweeper."

Everything a reader needs to understand about Blake's book is all there on the title page: *Songs of Innocence*, printed in 1789 by a W. Blake, who is also the author. That should just about do it, right?

Wrong. Without the title page in its original trappings, the text communicates only part of the message. For example, many of the themes inside *Songs of Innocence* deal with safeguarding children, birds, old people, and other indelibly cute creatures from the evils of the world. Take a look at the original title page illustration on the next page (a Blake relief etching, by the way), and you'll see that the narrative shifts a bit.

First, we feel a moral imperative to apologize for reproducing that title page in black and white. It is simply breathtaking in its original state, with a delicate wash of blues, reds, greens, and yellows. You're missing something extraordinary if you don't see it in color, an experience so beautiful that it's almost painful to tease you with this echo. But that is the point. When the illustration is translated into this medium, part of the experience is lost. Still, as bad as that is, you're missing way more when you have only the text in front

Individual title page of *Songs of Innocence* from *Songs of Innocence and Experience*, by William Blake, printed in 1795. *Image courtesy Yale Center for British Art.*

of you, which is how Blake's poetry is most commonly reproduced, and how most of us first read him.

When the text is combined with the image as Blake intended, we notice something else going on, something a bit at odds with the title. Blake specialist Robert Essick explains, "An adult reading to or lecturing children is a common motif in frontispieces and title-page vignettes of eighteenth century children's books. Blake exploits this convention by making one significant revision: the heavily dressed mother or nurse *shows* the book . . . to the boy and girl" [emphasis added].

It's as if Blake were somehow condoning the idea of children reading (and one of them a girl, no less!). By now it shouldn't surprise you that reading in the eighteenth century was a seriously dangerous pastime. Historian Robert Altick summarized the stakes: "Popular education struck horror into the souls of those committed to eighteenth-century social theory. It promised sloth, debauchery, and the assumption of superior airs on the part of the people—followed, as the day the night, by irreligion and revolution."

And how about the tree that's looming above those little revolutionaries? The vine winding around its trunk is serpentine in nature, and there are strange apples straining against their boughs, leaning forward like delicious orbs of good and evil hovering tantalizingly above youthful innocence. What seems at first like a totally mundane title page for a kid's book becomes a perilous scene of biblical proportions that makes you want to scream out, *Don't read it! Run away! The call is coming from inside the book!* After seeing the words in their original context, we suddenly have to revise the interpretation that these poems are meant only for children—or that they're even innocent at all.

Another example from Blake's *Songs of Innocence* can be found in the poem "The Little Black Boy," in which Blake takes a swing at the racism and white entitlement of eighteenth-century England.

Read on its own, without the accompanying illustrations, the poem is about a little boy who notices he's different from other boys: "White as an angel is the English child: / But I am black as if bereav'd of light."

"And we are put on earth a little space," the boy's mother tells him, "That we may learn to bear the beams of love. / And these black bodies and this sun-burnt face / Is but a cloud . . ."

Totally fine with this, says every Englishman, because it's 1789.

The boy then continues, addressing the white child, "And thus I say to little English boy. / When I from black and he from white cloud free / . . . I'll shade him from the heat till he can bear, / To lean in joy upon our father's knee. / And then I'll stand and stroke his silver hair, / And be like him and he will then love me."

Sounds about right. This is what little black boys should aspire to be: shade for white people. To be loved by white people, you must serve white people. And stroke their silver hair. Still, add Blake's illustrations, and the story changes a bit. The last plate shows Jesus not with two children "free" from color, but with both a white *and* a black child at his knee.

As you can see on the following page, the little black boy is still a little black boy. Despite being in the presence of God, he hasn't shed his "cloud" of dark skin as promised. In fact, he even appears to have angelic wings by virtue of the . . . we're going to say lambs (?) feeding in the background. Or large, furry newts? Shaved ocelots, maybe? Whatever. In this copy, the black boy has wings and seems to be interceding on behalf of the white child, who has not yet learned to "bear the beams of love." As one scholar noted: "Thus the design puts [the white boy] in the position of the lost soul who has been rescued by his black 'guardian angel.'"

Books are not created in a vacuum. The larger context of a work, including material elements beyond the printed words, makes up an inextricable part of the interpretation. Blake's work is endlessly

"THE LITTLE BLACK BOY (CONTINUED)," FROM *SONGS OF INNOCENCE AND OF EXPERIENCE*, BY WILLIAM BLAKE, PRINTED IN 1795. *Image courtesy Yale Center for British Art.*

challenging, which is one reason it's so beautiful. As readers, we lose the fundamental Blakeness of a work—that weird, provocative, glorious spirit—the further we push it out of context.

It is easy to take these small details for granted, even though they can completely alter our experience of a book. For example, most of us don't normally notice how illustrations, merely by their placement, can change the meaning of texts. Consider the 1796 edition of *Leonora*, translated into English from the original German by J. T. Stanley. The publisher, William Miller, commissioned William Blake to create three original designs for the poem. One of those designs would be used as the frontispiece, opposite the title page. The frontispiece essentially sets the tone and expectations of the entire book, and is far and away the most important illustration. In many books published in the eighteenth century, it is the only illustration.

In the spirit of the Grimm brothers' fairy tales, which were being collected in Germany about the same time, *Leonora* is the story of a young woman whose life goes awry in grotesque ways. As the poem opens, she is awaiting the return of her fiancé, William, from the Seven Years' War (1754–63, and yes, we're aware that's nine years; historians aren't good at naming things. Or math, apparently). When Leonora's waiting proves fruitless, she blames God, much to the disapproval of her mother, who tries to console her by saying:

Who knows but that he yet survives
Perchance far off from hence he lives
And thinks no more of you.
Forget, forget the faithless youth
Away with grief, your sorrow soothe
Since William proves untrue.

Sometimes mothers know just what to say to mend a broken heart. This was definitely not one of those times. Telling a teenage girl not to worry because her true love isn't dead, he's probably just an asshole who has a family of his own by now, is the exact opposite of helpful. Not surprisingly, Leonora acts like a teenager, dramatically beating her breast and praying for death.

Her fiancé then suddenly shows up on horseback in front of her house and invites her on a midnight ride. But here's the catch: she has to go right now, and can't tell anyone. So many red flags to unpack here, but Leonora is young and doesn't realize how very awful German fairy tales are. After riding around talking about dead people for a while, William then reveals himself to be Death, who's come to take her to hell for blaspheming God. The earth then cracks open to receive her.

Despite the genuinely disturbing latter half of the poem, the various English translations of *Leonora* end on an optimistic note. Leonora exclaims, "Thy will be done / Lord, let thy anger cease," and "[t]he spectres vanish'd into air . . . all was hushed in peace." In one version, William even returns from the war revealing that the whole poem was just a dream from the mind of a feverishly lovesick girl.

Various illustrators created designs for the immensely popular *Leonora*, which spurred several translations, play adaptations, and musical compositions. On the following page is an original pen-and-ink drawing for a famous scene in *Leonora*, by the German painter Johann David Schubert, made a few years after Blake's commission.

You can almost see the Disney film in the making: the dashing hero with great hair, the doting maid whose body not only defies the physics of riding a horse, but does so in order to showcase her tiny waist, supple breasts, and cartoonish doe eyes. And is that a nude ankle we see?

"LEONORE," ORIGINAL PEN-AND-INK DRAWING BY JOHANN DAVID SCHUBERT, CIRCA 1800. *Image courtesy Schmidt Auctions.*

Blake inv:
Perry. sc:
O! how I dreamt of things impoſsible,
Of Death affecting Forms least like himself;
I've seen, or dreamt I saw the Tyrant dreſs,
Lay by his Horrors, and put on his Smiles;
Treacherous he came an unexpected Guest,
Nay, though invited by the loudest Calls
Of blind Imprudence, unexpected still;
And then, he dropt his Mask.
Alter'd from Young.

FRONTISPIECE OF THE 1796 STANLEY TRANSLATION OF *LEONORA*, DESIGNED BY WILLIAM BLAKE. *Image courtesy Christie's Auction*

Now look at what Blake did.

Good God, man! Leave something for our nightmares. And this was the frontispiece, the most important image in the book. Blake's illustration changes the focus of the entire text from a theme of eventual redemption and God's ceaseless love to a cacophony of muscular butts, limbs that aren't physically possible, and a teenage maiden who's got no goddamn time for flashing ankles.

In case you're wondering—no, this design didn't play well with readers. From the *British Critic*: "[Concerning] the distorted, absurd, and impossible monsters, exhibited in the frontispiece to Mr. Stanley's last edition . . . [Blake] substitutes deformity and extravagance for force and expression, and draws men and women without skins, with their joints all dislocated, or imaginary beings, which neither can nor ought to exist."

That illustration changed readers' perceptions of *Leonora*. Single images really can have that much power. Take Sherlock Holmes as a further example. When we picture the world's most famous detective, we see him sporting that iconic deerstalker hat. That fashion choice comes not from the author, Arthur Conan Doyle, but from an image by Sidney Paget, the man who illustrated the Sherlock Holmes stories when they first appeared in the literary magazine *The Strand*.

It's worth noting that critics of Blake's *Leonora* weren't being entirely fair in their judgment. After all, the poem *was* an eighteenth-century German tale in the vein of birds pecking out the eyes of Cinderella's stepsisters. When Death reveals himself to Leonora at the end of the poem, it goes like this:

Scarce had he spoke, when, dire to tell
His flesh like touchwood from him fell
His eyes forsook his head.
A skull, and naked bones alone

Supply the place of William gone
'Twas Death that clasp'd the maid.

Now off to bed, kiddies. Blow out the candles. Sweet dreams. Blake's frontispiece is pretty faithful, if you ask us. (We'll also take this opportunity to emphasize how much we love this image. There's something truly breathtaking about Blake's species of nightmare.) His contemporaries, however, did not agree. The biographer G. E. Bentley notes that after this publication, "most books with Blake's engravings after his own designs were commissioned chiefly by his close friends . . . or were published by himself." Blake's frontispiece to *Leonora* altered the focus of the work, and it was not a change appreciated by readers of his day.

Blake could have taken special umbrage at one line from the *British Critic*: namely, that the figures in his design portrayed "imaginary beings, which neither can nor ought to exist." He probably read that in a room somewhere and shook his head. Because, for William Blake, the figures from his illustrations weren't merely wisps of some roiling fever. They were actual beings that wove together the fabric of his reality.

WILLIAM BLAKE was born in 1757 to the owners of a hosiery shop in London. Even as a young boy, he seemed more interested in doodling on handbills than winding tape, wrapping parcels, and measuring yards of fabric. Also at an early age, God scared the shit out of him by peeping into his window. "You know, dear, the first time you saw God," his wife reminded him during one interview, "was when you were four years old, and he put his head to the window, and set you a-screaming."

God later tried to make up for this poor first impression by granting glorious visions to the young lad: such as the biblical prophet Ezekiel sitting under a tree watching him; or another tree filled with angels, who were, also, silently watching him; then there were the haymakers working in the fields surrounded by angelic beings, and *they* were all watching him. From an early age, Blake understood that God was watching you. All the time.

Blake was one of those people for whom the term *enthusiasm*, from a Greek word meaning "inspired by a god," truly applied. "When [Blake] said 'my visions,'" wrote journalist and confidant Henry Robinson, "it was in the ordinary unemphatic tone in which we speak of everyday matters."

Many people who spent time with Blake became convinced that he was suffering from delusions, or was outright insane. One of the harshest critics of his work railed, "When the ebullitions of a distempered brain are mistaken for the sallies of genius . . . the malady has indeed attained a pernicious height." He went on to call Blake "an unfortunate lunatic, whose personal inoffensiveness secures him from confinement."

Another close friend of Blake's put it in the most politically correct way possible: "I very much fear his abstract habits are . . . much at variance with the usual modes of human life." And Blake's wife famously remarked, "I have very little of Mr. Blake's company; he is always in Paradise."

Blake genuinely believed he interacted with ghosts, angels, and demons on a daily basis. He often awoke to "harps which I hear before the Sun's rising." And, okay, one of his poems was dictated to him by a faerie in his parlor. But does that mean Blake was suffering from psychosis? Is that how we judge a person's sanity? By the medical definition of "psychosis," yes, it is. But as Blake observed, "there are probably men shut up as mad in Bedlam, who are not so . . . possibly the madmen outside have shut up the sane people." And that is totally not something a madman would say. . . .

Blake's "religious convictions . . . [have] brought on him the credit of being an absolute lunatic," wrote Robinson, "this belief of our artist's in the intercourse which . . . he enjoys with the spiritual world has more than anything else injured his reputation." Robinson's assessment was probably accurate. Blake was not fully appreciated in his time, and he died nearly penniless. There's no doubt that his visions made for awkward social situations. Take this comment made to a small group of friends one evening: "I can look at a knot in a piece of wood till I am frightened at it." This creates a striking image of Blake as a wide-eyed old man who is constantly afraid of everyday objects within his cone of vision.

What Robinson overlooks in his assessment, however, is that much of Blake's brilliance was positively defined by his "intercourse" with the supernatural. As Doris Lessing has written, "Posterity, it seems, has to soften and make respectable, smooth and polish, unable to see that the rough, the raw, the discordant, may be the source and nurse of creativity." It was as if Blake's senses perceived more than the natural world around him—experiences and information lost to the rest of us.

This "talent" manifested itself from his very first venture into the printing houses of London. Admitting that his son would never be great at sewing and dressmaking like a proper haberdasher, Blake's father enrolled him in the drawing school of Henry Pars from ages ten to fourteen. Upon graduation, young William was taken to interview as a potential apprentice to William Wynne Ryland, an artist and print seller who once served as engraver to the king. The fee that Ryland would have charged was probably hefty (somewhere in the neighborhood of one hundred pounds sterling), but Blake's father was willing to pay if his son's future could be comfortably secured.

The interview did not go well. "Father," Blake reportedly said, "I do not like the man's face: it looks as if he will live to be hanged."

On the list of things not to say during a job interview, this has to be near the top. You are suggesting that your potential employer (a) Looks like he might be a criminal; or (b) Is ugly enough that someone *should* hang him. In Mr. Ryland's case, however, it turns out that *a* was the correct answer. Falling on hard times, William Ryland was found guilty of forging bills from the East India Company to the tune of £7,114. In August 1783 (eleven years after Blake's interview), he was indeed hanged from a tree.

Blake's father took him next to see James Basire, the principal engraver to both the Royal Society and the Society of Antiquaries. Since Basire possessed a face that didn't suggest future crimes, Blake was apprenticed to him for seven years, at the cost of fifty-two pounds. From Basire, Blake learned the art of engraving, which included, among other things, cutting clean, straight lines into copper plates; cutting angled lines to represent shade (cross-hatching); using a "dot-and-lozenge" technique, which added small dots between the cut lines to create a 3D-esque depth of field; and the specialty of Basire's shop, intricate, curving "worm-lines."

Blake also learned how to etch copper plates with aqua fortis, or nitric acid. First coating the entire plate with acid-resistant wax, an engraver would sketch out the illustrator's design with a needle. Next, he would drop the plate into an acid bath, which bit those traced lines into the metal, but left intact the parts of the plate covered in wax. Blake later explained this process to one of his own apprentices in a way that brings to mind Patrick Swayze and Demi Moore over that potter's wheel in *Ghost*. "Take a cake of virgins wax . . . and stroke it regularly over the surface of a warm plate (the plate must be warm enough to melt the wax as it passes over) then immediately draw a feather over it and . . ." Good Lord, Mr. Blake, you are about to exceed the minimum age requirement for this book!

After seven respectable years with Basire, Blake graduated to journeyman, and was likely sent off with the customary gift of a "double suit of clothes and the tools of his trade." Rather than open his own studio, however, Blake felt the inspiration of God instructing him to return to art school. So it was that he found himself at the Royal Academy in Somerset House in 1779. He was a capable but obstinate student, unwilling to bend to the demands of instructors with whom he disagreed. According to his preeminent biographer, Blake "distrusted oil color as a modern and perverse invention." So what would he consider a more established and hallowed medium? Watercolors. Obviously.

Watercolors come with their own problems. They fade easily in the sun and aren't nearly as permanent in application as oil. Blake wanted to get around this, and fortunately, God showed him how—or, rather, the father of God showed him how.

"[Blake] ground and mixed his water-colors himself on a piece of statuary marble, after a method of his own, with common carpenter's glue diluted, which he had found . . . to be a good binder. Joseph, the sacred carpenter, had appeared in vision and revealed *that* secret to him." Sure, seeing dead people means you get branded a lunatic by your closest confidants, but it also means you get free art supply recipes. Fair trade.

After leaving the Royal Academy, Blake married, and took charge of his younger brother Robert after their father passed away. With the inheritance money, he was able to purchase an engraver's printing press, a big rolling press that was capable of producing the much greater pressure required for common intaglio illustrations. This came in handy when he and a partner opened a print shop in 1784. Because they were both trained engravers, they didn't have to hire others to create the material sold in their shop. Keeping everything in house cut down on the costs of production.

Life seemed to be going well for William Blake. His business was successful enough; he had a competent and loving spouse who helped in the shop; and he was even training his beloved brother Robert to follow in his footsteps as an engraver. Three years later, however, the scene was markedly changed. Robert lay coughing up blood, deep in the last stages of tuberculosis; William barely moved from his bedside during the last two weeks. When Robert died at the age of twenty-four, Blake said he saw the young man's spirit ascend to heaven, clapping for joy. As happy as Robert's ghost may have been, the brother he left behind cracked inside. He collapsed from exhaustion and for three days could not be roused, not even to attend Robert's funeral.

More than a decade later, Blake found solace in his ability to commune freely with Robert. "Thirteen years ago I lost a brother," he wrote, "and with his spirit I converse daily and hourly . . . I hear his advice and even now write from his dictate." During one of those conversations, Robert's ghost revealed the secrets of relief etching—because, apparently, when you die you become much better at your occupation than you ever were in life.

For Blake, the ability to etch in relief (to paint directly onto a copper plate and simply melt away the surrounding copper) commenced a rebirth in the art of illuminated manuscripts. Illuminated manuscripts added hand-drawn ornaments, borders, and miniature paintings to embellish and beautify a text. During most of the first three hundred fifty years of print, graphic elements were usually added to books by what could be carved into a woodblock or scratched onto a metal plate. Now, by painting directly onto the medium, Blake could incorporate styles of art that were largely foreign to the mechanical printing process. The resulting books are an ineffable joy to behold, and a chance to see one of them a major event. The last copy of *Songs of Innocence* to sell at auction, for example, went for almost a million dollars in 2001.

In a 1793 advertisement for his relief-etched books, Blake wrote that his new style of printing was "more ornamental, uniform, and grand, than any before discovered, while it produces works at less than one fourth of the expense . . . The illuminated books are printed in colors, and on the most beautiful wove paper that could be procured." It was "a method of printing which combines the painter and the poet," making the author, the artist, the engraver, and the printer all the same person.

After the invention of his relief etching, Blake finally achieved complete creative control over his work, with no middleman inserting his own interpretation. In many books, this species of seemingly harmless meddling can be as simple as changing the font, or the layout of the text on the page, or the type of paper. It's not unusual for an author to try to maintain control of these elements as well. For instance, in the twentieth century, James Joyce wrote the story of *Ulysses* as a single day in a leap year, and he made sure that the first edition mirrored this by having exactly 366 leaves.

William Blake wanted every choice that affected the final product to be his alone. As a result, when we use an edition that reproduces only the text (or an illustrated edition without the color), we lose some of Blake's consciously crafted meaning. Consider the way in which the painter and poet work against each other in Blake's "London," from *Songs of Experience*. "London" is a cheery poem about life in a stand-up town where the Janes and the Darbs beat their gums on the street corners and laugh—a town where just drawing one's breath is the cat's meow. Just kidding, it's about how much London, as a place for human beings to live, sucks. "In every face [you] meet," there are "marks of woe." There are "mind-forg'd manacles." "[Y]outhful Harlots" curse. "[T]he hapless Soldier's sigh / Runs in blood down Palace walls." It's the eighteenth-century equivalent of a half-star Yelp review.

LONDON

I wander thro' each charter'd street,
Near where the charter'd Thames does flow
And mark in every face I meet
Marks of weakness, marks of woe.

In every cry of every Man,
In every Infants cry of fear,
In every voice: in every ban,
The mind-forg'd manacles I hear

How the Chimney-sweepers cry
Every blackning Church appalls,
And the hapless Soldiers sigh
Runs in blood down Palace walls

But most thro' midnight streets I hear
How the youthful Harlots curse
Blasts the new born Infants tear
And blights with plagues the Marriage hearse

"LONDON," FROM *SONGS OF INNOCENCE AND OF EXPERIENCE*, BY WILLIAM BLAKE, PRINTED IN 1795. *Image courtesy Yale Center for British Art.*

Yet, this is the original illustrated page of "London." Above the manically depressive text stands the image of a young boy helping a crippled old man, an illustrative act of charity overshadowing the textual human woes. Unless that kid is leading the old man to a gang of street punks who beat him up and steal those sweet crutches, the illustration and the poem are sending two different messages to the reader that must be considered together when experiencing "London." Taken as a composite work of art, the poem can be read as a question and answer: the text as the problem and the illustration as the solution.

The additional context of Blake's illustrations changes how we read the text, but important meanings can be imposed by other factors we may not immediately notice. When Charles Dickens was publishing his serial novel *Hard Times* in 1854, he did so within the pages of his own weekly magazine, *Household Words*. The weekly also contained nonfiction articles, but their arrangement in relation to the fictional *Hard Times* installments was designed to provide setting and context to the events of the novel. Actual news articles about the rights of workers, failed safety regulations, accidents, and industrial reform coincided with the lives and deaths of Dickens's fictional characters. When reading *Hard Times* as a standalone novel separate from *Household Words*, we miss major components of the argument Dickens is making.

As the scholar Jerome McGann argues, the words on a page offer only "the merest glimpse of that complex world we call literary work and the meanings it produces." William Blake is an exquisite example of this concept because his "composite art" of word and image allows for unusual complexity, with layers upon layers of meaning.

Sometime just before 1794, William Blake had his most powerful vision. A godlike figure appeared to him and "hovered over his head at the top of his staircase." What God was doing up there is anyone's guess (though, if Blake's experiences as a youth are any indication, God was probably waiting to pop out and scare the bejesus out

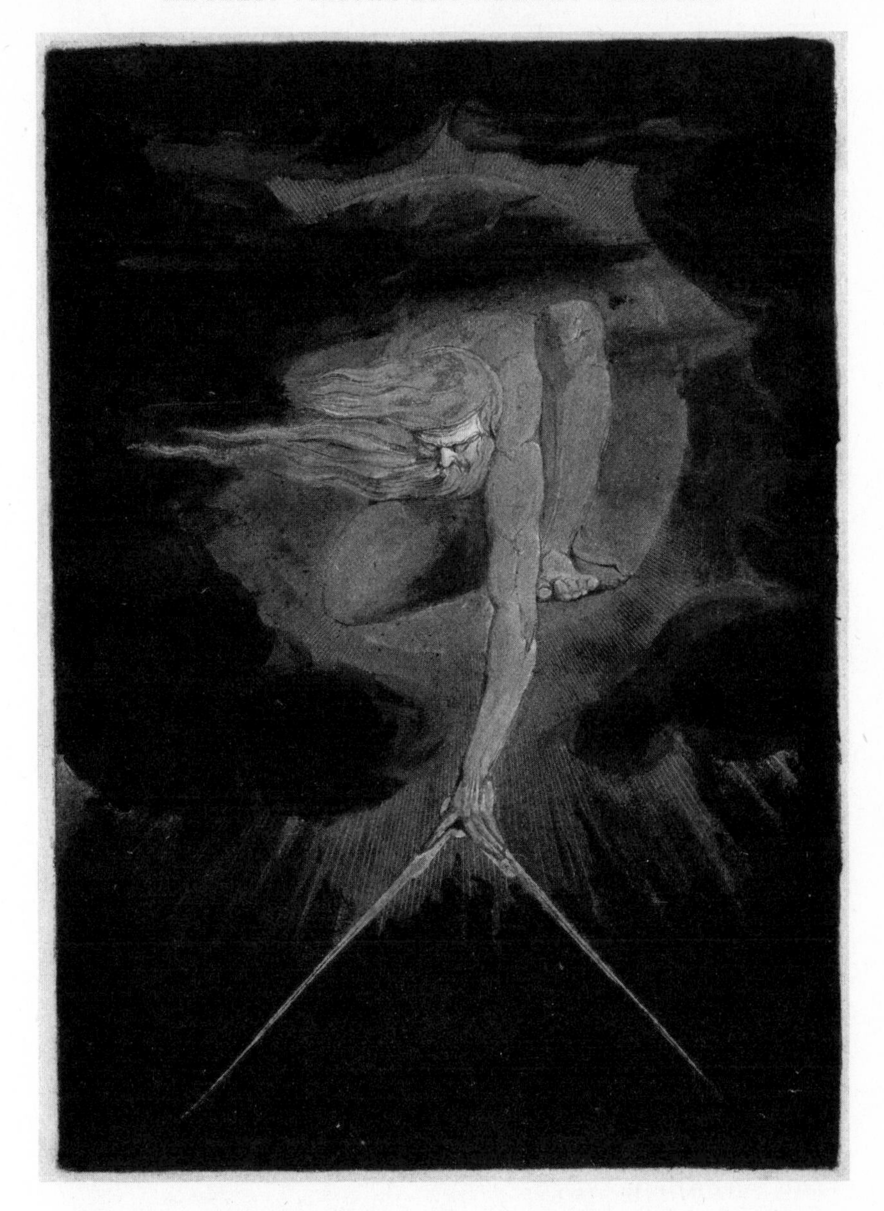

EUROPE A PROPHECY FRONTISPIECE BY WILLIAM BLAKE, PRINTED IN 1795. *Image courtesy Yale Center for British Art.*

of him). Upon seeing this vision, Blake felt inspired to create one of his most famous illustrations, "The Ancient of Days."

EUROPE A PROPHECY TITLE PAGE BY WILLIAM BLAKE, PRINTED IN 1795. *Image courtesy Yale Center for British Art.*

This design became the frontispiece of Blake's lengthy poem *Europe a Prophecy* (1794). Filled with all the doom and gloom and apocalyptic fervor that the 1790s could muster (which was a surprisingly large amount), *Europe*'s text depicts a world of "almost hopeless torment," where understanding the causes of evil does nothing to free mankind from their collective suffering. Just a light read, really.

Next to "Ancient of Days" sits *Europe*'s title page, shown above.

Not only the illustrations, but their specific placement in the book, tell a story. The creator God from Blake's frontispiece reaches down from the sky to fashion not a garden of Edenic beauty, or a naked man and woman, or even dinosaurs who need to hurry up and go extinct before the naked man and woman get booted from the magical garden for fruitful translations. Instead, reading the book's illustrations from left to right, God first creates a traditional symbol of evil: the Serpent.

We tend to assume that a book's narrative is communicated simply by the words alone. That's an obvious mistake when you consider Blake's work, but the principle applies much more broadly. Some modern authors, experimenting with the new medium of digital-born texts, have created narratives that change based on how you interact with them. In Shelley Jackson's 1996 hypertext novel, *Patchwork Girl*, the reader is presented with the body of a woman. Each body part has its own individual backstory. "I am buried here. You can resurrect me, but only piecemeal. If you want to see the whole, you will have to sew me together yourself." A woman's body, graveyards, journals, and even a quilt—the reader must click on various hyperlinks, embarking on a tale of a female Frankenstein's monster created interactively by the reader. The way a reader advances the story "characterize[s] a good deal of the way we conceive of gender and identity." As brilliantly crafted as Jackson's narrative is, it could not exist outside its original medium without dramatic loss to its meaning.

For William Blake, certain symbols were bursting with significance. The Serpent, like that from *Europe a Prophecy*, held an interesting place in Blake's mythos: "Satan . . . is father & God of this world." And there was this gem: "The Prince of darkness is a gentleman & not a man[;] he is a Lord Chancellor." All right, Mr. Blake. How did you come upon this information?

"For many years . . . I longed to see Satan."

No story that starts this way ever ends well.

"I never could have believed that [Satan] was the vulgar fiend which our legends represent him . . . At last I saw him. I was going upstairs in the dark, when suddenly a light came streaming amongst my feet, I turned round, and there he was looking fiercely at me through the iron grating of my staircase window." Blake really needed to stay away from windows and staircases at night.

". . . through the bars glared the most frightful phantom that ever man imagined. Its eyes were large and like live coals—its teeth as long as those of a harrow [spiked plough], and the claws seemed such as might appear in [a] distempered dream . . . It is the gothic fiend of legends, the true devil—all else is apocryphal."

To recap: I wanted to meet Satan. I didn't believe he was as awful as the legends say. There was a light at my window. Whoa! Just kidding about that Lord Chancellor nonsense; he is a monster! A goddamned monster!

Blake's demons had many faces. "The Tyger," arguably his most famous poem, published in 1794, starts off like a children's rhyme. Then it quickly turns into something an angsty teenager might read to summon Cthugha, the Lovecraftian fire god.

Tyger Tyger, burning bright,
In the forests of the night;
What immortal hand or eye
Could frame thy fearful symmetry?

And what shoulder, & what art,
Could twist the sinews of thy heart?
And when thy heart began to beat,
What dread hand? & what dread feet?

What the hammer? what the chain,
In what furnace was thy brain?
What the anvil? what dread grasp
Dare its deadly terrors clasp!

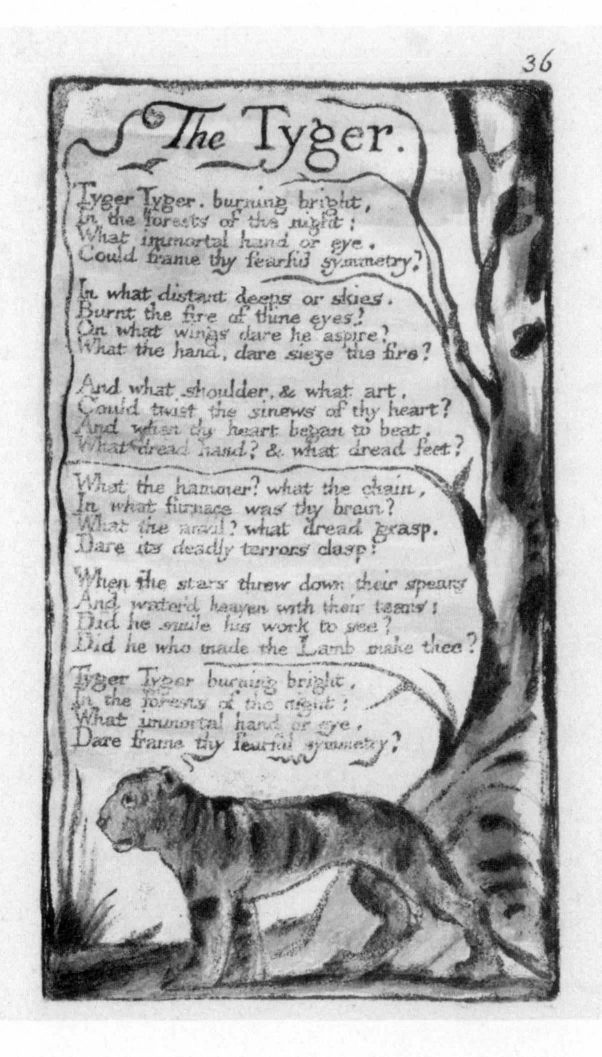
36

The Tyger.

Tyger Tyger. burning bright,
In the forests of the night:
What immortal hand or eye.
Could frame thy fearful symmetry?

In what distant deeps or skies.
Burnt the fire of thine eyes!
On what wings dare he aspire?
What the hand, dare sieze the fire?

And what shoulder, & what art,
Could twist the sinews of thy heart?
And when thy heart began to beat,
What dread hand? & what dread feet?

What the hammer? what the chain,
In what furnace was thy brain?
What the anvil? what dread grasp.
Dare its deadly terrors clasp?

When the stars threw down their spears
And water'd heaven with their tears:
Did he smile his work to see?
Did he who made the Lamb make thee?

Tyger Tyger burning bright,
In the forests of the night:
What immortal hand or eye,
Dare frame thy fearful symmetry?

"THE TYGER," IN *SONGS OF INNOCENCE AND OF EXPERIENCE*, BY WILLIAM BLAKE, PRINTED IN 1795. *Image courtesy Yale Center for British Art.*

The text itself should be read only to children who *don't* cry at eighteenth-century German fairy tales—but the design Blake chose to personify this hellish beast of youthful nightmare looks like this:

Well, that's not very frightening at all. It's kind of adorable. Less

"GHOST OF A FLEA," ORIGINAL WATERCOLOR BY WILLIAM BLAKE, C. 1819–20. *Image courtesy Tate Museum.*

"dare its deadly terrors clasp" and more "I CAN HAS DEADLY TERRORS?"

Blake's work can often be more intriguing for its questions than its answers. Why would he draw his forest demon like that? Who knows? The weird tree face at the top is scarier than the feline predator. It wasn't as if William Blake were incapable of designing "distorted, absurd, and impossible monsters." Recall his illustration of *Leonora* or, better yet, this terror of Christian indecency.

This is the "Ghost of a Flea," circa 1819. And here is how this amazing image came about:

> I [Allan Cunningham] called on him one evening, and found Blake more than usually excited. He told me he had seen a wonderful thing—the ghost of a flea!
>
> "And did you make a drawing of him?" I inquired.
>
> "No, indeed," said he, "I wish I had, but I shall if he appears again!"
>
> He looked earnestly into a corner of the room, and then said, "here he is—reach me my things—I shall keep my eye on him. There he comes! His eager tongue whisking out of his mouth, a cup in his hand to hold blood, and covered with a scaly skin of gold and green";—as he described him so he drew him.

The last sentence of Cunningham's description is priceless. You walk into your friend's house, and he tells you that he's seen the ghost of a flea. Generally, being familiar with the size and shape of a flea, you ask him to draw it, because—why not? It will probably be cute. He scribbles for a few minutes and then hands you the picture of a six-foot-two bipedal fish monster carrying a mixing bowl of human blood. When Cunningham wrote, "[A]s he described him so he drew him," he nonchalantly captured one of the most fundamental aspects of his friend's world. For Blake, the grotesque and the heavenly were mixed with the most ordinary features of life.

William Blake was a visionary in every sense of the word. While regarded mostly as an oddity during his lifetime—whenever he went missing for a while, people assumed he was either dead or finally locked up in a madhouse—he invented an illustration method that would be admired and remarked upon for centuries to come. In fact, his precise relief etching technique wasn't fully understood until the second half of the twentieth century. But while his technical achievements are interesting and beautiful, the intentions that drove him to these breakthroughs are equally compelling.

Blake couldn't bear working only as an engraver, accomplishing the "mere drudgery" of imitating some other artist's original work. He took control of the medium, forcing readers to recognize that a text should not be analyzed apart from its materiality. The bibliographer G. Thomas Tanselle says, "Books are a part of material culture. Every artifact, every physical object made by human beings, is a record of human effort at a particular time and place, as well as a tangible link to all the succeeding moments of its life." Blake understood this, and his printing methods were meant to be an integral part of his message. He explains this in a way so very Blakean: "The notion that man has a body distinct from his soul, is to be expunged; this I shall do, by printing in the infernal method, by corrosives [etching], which in Hell are salutary and medicinal, melting apparent surfaces away, and displaying the infinite which was hid."

During the last years of his life, Blake slipped into acute poverty, and varying degrees of mental wellness. Some who met him described him as "so evidently insane, that the predominant feeling in conversing with him, or even looking at him, could only be sorrow and compassion." At other times it was said of him, "I remember William Blake, in the quiet consistency of his daily life, as one of the sanest, if not the most thoroughly sane man I have ever known." Which is sweet, but also hints of protesting a bit too much. *How's your uncle, Bill? Oh, fine, fine. Completely sane. Really, he's great. Not insane at all.*

Henry Robinson, the journalist from whom comes a treasure trove of Blake interviews, diagnosed Blake in 1810 with "monomania," a nineteenth-century psychiatric term for a person whose mind was mostly sound, except for one pathological preoccupation. From Robinson's interviews we get rare (and sometimes heartbreaking) insights into Blake's progressing preoccupation.

"He is now old," Robinson wrote, "pale, with a Socratic coun-

"THE CIRCLE OF THE LUSTFUL, PAOLO AND FRANCESCA," ORIGINAL ILLUSTRATION TO DANTE'S *INFERNO*, BY WILLIAM BLAKE. ENGRAVING WITH DRYPOINT, 1826–27. *Image courtesy John Windle, Antiquarian Bookseller.*

tenance, and an expression of great sweetness, though with something of languor about it except when animated, and then he has about him an air of inspiration . . . [he] spoke of his paintings as being what he had seen in his visions . . . In the same tone he said repeatedly, 'the Spirit told me.' . . .

"[Blake] paused & added, 'I was Socrates,' and then, as if correcting himself, said, 'a sort of brother. I must have had conversations with him. So I had with Jesus Christ. I have an obscure recollection of having been with both of them.'"

To generate income and keep him busy during the last months of his life, Blake was commissioned by one of his friends to design engravings for a new edition of Dante's *Inferno*. Never content with a simple illustration of another's work, he used the medium to make his own commentary on Dante's epic, described by one scholar as "Blake's most drastic act of reinterpretation." Blake died before the series was completed, but the images left to us have proved to be "triumphs of the engraver's art, among the finest line-engravings ever made."

William Blake died in 1827, at the age of sixty-nine. The exact causes of his death are unknown, but there is evidence that he was suffering from sclerosing cholangitis, an inflammatory bowel disease that may have been "caused or aggravated by chronic copper intoxication." If that's true, then, by his art, he lived, he dreamed, and he died.

Blake was a man of contradictions. He was haunted by ghosts and demons, yet from those interactions, he created some of the most alluring and innovative works of the era. As one biographer notes, William Blake lived in a paradise of his own making, in "realms of gold." This is a place "we too may dwell . . . if we use our imaginations"—if, that is, we can stomach the leering gods, demons, and terrifying man-fleas. Perhaps to join Blake, all we really need to do is listen. *Escuchar a los muertos con los ojos*, as Francisco de Quevedo said. "Listen to the dead with your eyes."

8

THE MEMOIR THAT KILLED HER MEMORY

IN LATE OCTOBER 1795, ONE of the most revolutionary authors of the last decade poised to throw herself from London's Putney Bridge. Rationally, the author made this decision because she no longer believed in the superstitious tales of eternal punishment promised to those who "self-murder." Irrationally, the author made this decision because she had been jilted by a lover. In a note calculated to haunt him, she wrote, "May you never know by experience what you have made me endure . . . in the midst of business and sensual pleasure, I shall appear before you, the victim of your deviation from rectitude." That is a serious curse: *Every time you make a business deal, I will be there silently watching. Every time you bed another woman, I will be there silently watching. Not so easy to get that pen of yours up now, is it?* The writer of this odd suicide note was none other than Mary Wollstonecraft, the famous author of the protofeminist treatise *A Vindication of the Rights of Woman*.

Even before the invention of the printing press, writers were wrestling with the question of professional authorship. Could writing be viewed as an actual career? Of course it can, says anyone who's read *Harry Potter*. But making a living from writing is a fairly new phenomenon in the history of print. For the first three hundred fifty years, "professional author" was a made-up career, like philosopher, fantasy football manager, or American vice president.

Over the course of the 1700s, the trade in printed books expanded so dramatically that, for the first time, many authors could actually list "writing" as their primary occupation. The number of titles available in England jumped 50 percent, from 1750 to 1775, but leapt 170 percent from 1775 to 1800. Writers enjoyed more venues to publish reviews, essays, histories, and even those newfangled and morally bankrupt "novels" that everyone was warning about. Hit the Zeitgeist with a single work, and your name could be lauded all across Europe.

At this same time, a philosophical shift was creeping into the world of print. Authors were starting to be celebrated as heroes, special voices of authority sustained by virtue of their unique genius. They even began asserting, and winning, rights to previously nebulous concepts such as "intellectual property," through landmark court cases such as *Pope v. Curll* (1741). This all sounds great, until you remember that this is the eighteenth century, a time before proper sewage systems, when bloodletting was still considered an important treatment for illness, and when married women were not considered fit to own property. Mary Wollstonecraft took on the challenge of becoming a literary authority despite the impropriety of possessing a vagina. As a result, Wollstonecraft's reputation has varied so wildly over the past two hundred years that you're just as likely to find statements calling her the "Mother of Feminism" as a "Prostitute."

When Wollstonecraft decided to end her life in 1795, October on the River Thames would have been about the worst time to do it. Autumn was apparently the busiest season for desperate women throwing themselves into a watery grave. According to her biographer Janet Todd: "Unconscious, [Wollstonecraft] floated downstream, until pulled out of the river by fishermen, doubtless used to suicides . . . The Royale Humane Society had been set up to pursue the enlightened policy of thwarting self-murders by receiving, and if possible, resuscitating bodies found floating in the Thames."

How would a lovesick suicide attempt affect the reputation of the fierce author of *A Vindication of the Rights of Woman*? Here's one report from *The Gentleman's Magazine* just after Wollstonecraft was pulled from the icy river. "From a mind of such *boasted* strength we naturally expect fortitude; but, in this instance, she was weak as the weakest girl" [emphasis original]. So, only authors with the weakest, girliest of minds commit suicide. Just ask Ernest Hemingway.

It wasn't the suicide attempt(s) that would most damage Wollstonecraft's hard-earned reputation, however. It was a memoir. Written by the man she married a year and a half after the fateful Putney jump. William Godwin intended to memorialize his well-known wife to the annals of literary glory, but, in the end, he served only to relegate her to the trash heap of history.

MARY WOLLSTONECRAFT was born in 1759 to an unfortunate London family. By her own accounts, her father was a lazy tyrant who spent more time abusing her mother than supporting her. Her mother rarely pushed back; Wollstonecraft remembered her mother as more a slave than a spouse.

Ducking the rage of her frequently drunken husband, Wollstonecraft's mother took solace in her children—well, one of them, anyway, the eldest son, Ned. Generally neglected by both parents, Mary felt compelled to look elsewhere for acceptance. She found it in the London-based family of Matthew and Caroline Blood. Their daughter Frances (who went by Fanny, by the way. Fanny Blood.

Aaaand moving on) would become one of Wollstonecraft's closest friends and confidantes.

At twenty-four years old, Mary, two of her sisters, and Fanny opened schools for young women in Islington and Newington Green in an attempt to carve out financial independence. Schools at the time were about as stable as dot-com businesses after 2001. The start-up was cheap and the qualifications for employment low, but most ventures sagged under starving profits, and eventually went under. Within two and a half years, both Wollstonecraft's schools had failed.

Ironically, while Wollstonecraft was failing at running her schools, she was succeeding in becoming an authority on them. In the same year that her Newington Green school closed, she began work on her first book, *Thoughts on the Education of Daughters* (1787). She became a nationally recognized "pedagogue with her own theory of education." As the old saying goes, those who can, do; those who can't, teach; and those who can do neither, write a book about it . . .

Mary Wollstonecraft wasn't destined to become the eighteenth-century Danielle Steel. She was more in the neighborhood of a poverty-stricken single mom who writes in cafés and makes ends meet by doing translation work. And yes, that's an almost word-for-word description of J. K. Rowling, but Wollstonecraft didn't have *Harry Potter* in her back pocket. She had the YA book *Original Stories from Real Life; with Conversations Calculated to Regulate the Affections, and Form the Mind to Truth and Goodness.* Wollstonecraft worked for years as a translator, journalist, and editor to financially support herself.

The career path before Wollstonecraft was not easy. The concept of the professional author, one who could live solely off his writing, would have seemed more myth than reality to people living in the first centuries of print. For women, the idea was even more far-fetched. While there are some rock star exceptions, such as the fourteenth-century Venetian Christine de Pizan (who was widowed at twenty-five and made a living for herself and her three children as a popular author), until the eighteenth-century the world of print was guarded in a kind of aristocratic system, which generally excluded both the uneducated (see: poor) and the mentally inferior (see: women).

Most readers today take for granted the concept of the author, but in the thousands of years that humanity has been writing, we've had a hard time figuring out just what an "author" is. Plato argued that most poets should be cast out of the city as useless liars. In the best cases, authors were seen as *auctors*, tasked with creatively standing in the place of God. To take sole credit for one's work would have been akin to plagiarizing God. According to Dante, those guilty of fraudulent acts were sent to the eighth circle of hell, where they would be dipped in excrement for eternity.

Through most of the Middle Ages, writing wasn't even considered a paying gig: how can you sell something that was a gift from God? You can't. Just like you can't sell that pair of scratchy socks your nana knitted last year, you cannot, if you're the medieval writer Boethius, sell *The Consolation of Philosophy*. Both are gifts. Equal-magnitude gifts.

The concept of what an author did, what authority he had, and what he was owed for it, has never been consistent. The sixteenth-century humanist Erasmus was morally offended when someone suggested that his printer had paid him for his work. Even well into the 1700s, it's easy to find documentation that shows an author was "still being represented as just one of the numerous craftsmen involved in the production of a book—not superior to, but on par with other craftsmen." Yet, the blossoming of European capitalism rapidly expanded the market for books, which required more printers, more papermakers, more booksellers, more binders, and yes, even more writers.

With the trail blazed by the likes of Alexander Pope and Samuel

Johnson, Mary Wollstonecraft was one of a number of women who hoped she could earn economic independence on the same path. But because "authorship" overlaps "authority," it rubbed some gentlemen the wrong way that a woman could dare to acquire some of the latter. It's no coincidence that women writers have frequently been accused of plagiarism, from Anne Bradstreet in 1650 ("If what I do prove well, it won't advance, / They'll say it's stoln, or else it was by chance") to claims that Truman Capote was the actual writer of *To Kill a Mockingbird* in 1960. Nor is it surprising to see how many women writers, such as the Brontë sisters, used male pseudonyms when trying to establish themselves. For women, as one scholar put it, "authorship had to be denied so as to be attained." The golden age of the author was just beginning, and writers were required to check their coats and their womenfolk at the door.

Let's pretend for a moment that this period of print history is a recipe for a soufflé. We take 1700–1850 CE and whisk it up as hard as we can. Then we add three cups of money, seven tablespoons of Daniel Defoe (*Robinson Crusoe*), a dash of Samuel Richardson (*Pamela*), and, let's say, one hatful of Henry Fielding (*Tom Jones*). We're pretty sure that's how you make soufflés. You bite into this fluffy marvel of literary history, only to notice that something besides a basic familiarity with French cuisine is missing. Where's the Mary Brunton in this recipe? Where's the Charlotte Smith? Or the Sarah Fielding, Maria Edgeworth, and Elizabeth Montagu? Ann Radcliffe? Eliza Haywood? Maybe you added a little Jane Austen, so congratulations for that, but this is the moment when you realize that the soufflé you've just made is totally imbalanced. You put in way too much sausage and only one egg. (Good luck eating your next soufflé.)

The systematic excising of the contributions of women in this era has its own name. It's called the Great Forgetting. Mary Wollstonecraft's name has been penciled in and scribbled out of our soufflé recipe for two hundred years. Historian Seán Burke makes the point that "it would scarcely be an exaggeration to say that the struggles of feminism have been primarily a struggle for authorship." And you can bet this struggle is reflected in the world of rare book collecting. Many first editions by the female authors just listed can be purchased at a fraction of the price of those by their male counterparts—with the usual exception of Jane Austen, whose *Pride and Prejudice* can reach very high prices: seventy-five thousand dollars or more.

During the Romantic Age, readers were eager for more than just the products of the author's pen (gross). They wanted to know about the author himself. The *Confessions* of Jean-Jacques Rousseau, one of the greatest autobiographies ever written, was published at the end of the eighteenth century. So was the first "warts and all" biography: James Boswell's delightfully unflinching narrative of Samuel Johnson (the lexicographer behind the greatest English dictionary). Soon biographies started cropping up as a mainstay of nineteenth-century literary life. One of the most famous of these was Elizabeth Gaskell's *Life of Charlotte Brontë*, which praised Brontë's "great genius" (while simultaneously suggesting that her "duties" as a Victorian woman suffered for that genius). The first authoritative biography of Charles Dickens was written during this period, by his close friend and literary agent John Forster. But before that was the sexually charged, tell-all memoir of Mary Wollstonecraft by her grief-stricken widower—the one that obliterated her reputation for more than a century.

While the biographical approach to literary criticism can provide illuminating context to an author's work, it's also a double-edged sword. When sordid details of Wollstonecraft's personal life became public, critics finally found the weapon they needed to strip her of her authority.

Read Wollstonecraft's epitaph, and note the emphasis:

MARY WOLLSTONECRAFT GODWIN
Author of A Vindication
of the Rights of Woman
Born 27th April 1759
Died 10th September 1797

The crowning achievement of Wollstonecraft's literary career, commemorated in stone, was published in 1792, just five years before her untimely death. *Vindication* "proposed a model of what we would now call 'egalitarian' or 'liberal' feminism. Grounded in the affirmation of universal human rights endorsed by such Enlightenment thinkers as Voltaire, Rousseau, and John Locke, Wollstonecraft argued that females are in all the most important aspects the same as males, possessing the same souls, the same mental capacities, and thus the same human rights."

That Wollstonecraft had to start with, *Hey, everybody, men and women don't actually have different souls*, gives you an idea of what she was working with. Her treatise took a hard look at female education, arguing in particular that girls should be "educated in the same subjects and by [the] same method as boys." Latin, for example, was considered a cornerstone of a schoolboy's curriculum, whereas for girls, it made them "unmarriageable." Because the only thing worse than a woman talking back to you is a woman talking back to you in Latin.

What might Mary Wollstonecraft consider an "improper education"? Probably something like "the education of the women should be always relative to the men. To please, to be useful to us, to make us love and esteem them . . . to console us, to render our lives easy and agreeable: these are the duties of women at all times, and what they should be taught in their infancy." Cooking? Check. Mending holes in pantaloons? Check. But Latin? Hold on there, young lady, how exactly does *that* benefit a man?

The public appeal to separate male and female educations was made by none other than Jean-Jacques Rousseau in his 1762 treatise, *Émile, or On Education.* Rousseau was one of the most influential philosophers of the French Enlightenment, and *Émile* was, he claimed, the "best and most important of all my writings." Mary Wollstonecraft's *A Vindication of the Rights of Woman* was a giant *futue te ipsum* to Rousseau and the others who espoused his misogynistic views. In it she devotes a significant amount of time to dismantling Rousseau's ideas: "the mother, who wishes to give true dignity of character to her daughter, must, regardless of the sneers of ignorance, proceed on a plan diametrically opposite to that which Rousseau has recommended."

In addition to bettering female education, *Vindication* also advocated for joint possession of household resources, equal opportunity employment, women's suffrage, and equal pay for equal labor. Most of the core principles that would resurface in the gender equality movements of the twentieth and twenty-first centuries were established in print by Mary Wollstonecraft two hundred years before, earning her that nom de guerre "Mother of Feminism." But, in 1798, her husband published *Memoirs of the Author of A Vindication of the Rights of Woman,* which punted said *nom* into relative obscurity for the next century.

BY WRITING *Memoirs,* Godwin was paying tribute to Wollstonecraft, who had just died from complications of childbirth after delivering her second daughter. (This daughter, also named Mary, ran away with the poet Percy Bysshe Shelley in 1814. She would write *Frankenstein; or, The Modern Prometheus* four years later, aspects of which would falsely be credited to her more famous consort.) A radical himself and, at the time, an even more successful public intellectual, Godwin was deeply in love with his revolutionary wife. He respected her sharp intellect and writing skills. In the depths of his grief, he didn't seem to comprehend how others might read her works and not also fall in love with her. A candid biography of his extraordinary wife should endear her even more to the public, right? It worked for the infamously irascible Samuel Johnson, so why not Wollstonecraft? Her successes and failures would be immortalized in print, and in the end, her contemporaries would judge her from the vantage point of God, with all the fairness, mercy, and understanding requisite of such an office. Mary made mistakes, sure, but who hasn't? Unfortunately, for those who disagreed with Wollstonecraft's liberal, revolutionary writings, Godwin's *Memoirs* was a pile of fresh ammunition, just waiting to be loaded into the guns of their righteous indignation.

The woman that Godwin painted in *Memoirs* was an odd duck. From the start, she was a girl who, when beaten by her father, refused to submit. "The blows of her father . . . instead of humbling her, roused her indignation. Upon such occasions she . . . was apt to betray marks of contempt." *A young woman who will not yield to her father?*

"Dolls and the other amusements usually appropriated to female children, she held in contempt; and felt a much greater propensity to join in the active and hardy sports of her brothers . . ." *Uh oh. Nothing good can come of girls who act like boys.*

At nineteen, Wollstonecraft struck out on her own to pursue a career separate from her family. She seemed relatively uninterested in marriage, and was constantly striving to be an independent breadwinner. Young ladies who constantly follow after such masculine pursuits might be prone to other masculine undertakings. Like having sex whenever you feel like it and giving little thought to the aftermath . . .

(Ah, we are reading ahead in *Memoirs*. The difference, we might suggest, is that men who gallivanted through the countryside, leaving title aftermaths in their wake, were not professionally discredited for a hundred years as a result.)

"Unsex'd female" was the term popularized in this period to attack women such as Mary Wollstonecraft. While at first that sounds like a woman who just needs more sex to solve her problems, it actually meant a female who did "mannish" things, such as, apparently, having a career. When *A Vindication of the Rights of Woman* was published in 1792, at least one parent publicly denounced the treatise as infecting her daughters with this species of unsex'd mannishness:

"A 'mother' wrote to the *Ladies* [sic] *Monthly Museum* to lament that her four daughters had been corrupted by the book: one lost her 'softness' and indulged in horse-racing, fox-hunting and betting; a second had taken up Latin and Greek; a third was scientifically dissecting her pets; and a fourth was challenging men to duels."

By the mid-nineteenth century, Wollstonecraft's name was anathema. "It is to be lamented that Wollstonecraft, whom nature . . . meant should be a bright pattern of perfection to her sex, should, by her erroneous theories and false principles, have rendered herself instead, rather the beacon by which to warn the woman of similar endowments with herself, of the rocks upon which enthusiasm and imagination are too apt to wreck their possessor."

And that isn't even from some rabid Wollstonecraft critic. That was a selection from one of the earliest sympathetic discussions of her life, published in a *Literary Ladies of England* compilation from 1843, fifty years after her death.

For more than a century, and to some extent even today, Wollstonecraft couldn't escape the damage of her husband's ill-fated biography. Unlike a male author whose sexual dalliances could be revealed with his reputation relatively intact (Mr. Hemingway, again), Mary had no such luxury. A scarlet letter on her person equaled a scarlet letter on everything she had ever thought or written.

To understand the depth of her "immoral" actions, one needed only read *Memoirs* as far as the section on the Revolutionary War officer Gilbert Imlay. Captain Imlay was in his late thirties, tall, fit, and extremely handsome. These qualities could easily be assessed at a glance. But he was also a kind southern gentleman with considerable landholdings in his family state of Kentucky. Imlay's Revolutionary War veteran status lent him glamour in the tumultuous Europe of the 1790s. He was an American adventurer, a man of refinement, and a successful entrepreneur. He was the eighteenth century's Han Solo. How could you *not* fall for the guy? We're falling for him and we've never seen him before. Wollstonecraft developed the same feelings after traveling to Paris in 1793. Unlike some of her previous infatuations, Imlay was single, open-minded, and passionately interested in her.

Captain Imlay was also a confident man. Like, a very *con*(fident) *man*. Despite his filling Mary's head with visions of retiring to an idyllic farm in Kentucky, it's unlikely Captain Imlay could ever have returned to the Bluegrass State. For one, Gilbert Imlay was born and bred in New Jersey. He served briefly in the War for Independence, receiving the rank of captain. From whence Kentucky, you might ask? After the war, Imlay used his veteran benefits to lay claim to lands in Kentucky that had been (cough) appropriated (cough, cough) from local Native American tribes. He became a deputy land surveyor and used his position to make shady business deals on the side. Whatever the extent of his scams, they backfired, and Imlay fled the American continent ahead of a barrage of lawsuits.

How much of this Wollstonecraft actually learned is open to speculation. She certainly knew that Imlay's current profession was "French smuggler." During the violent upheavals of the French Revolution, the British Navy blockaded the country. Imlay threw himself into overcoming that problem by running grain, iron, soap, and other supplies through the siege lines. Wollstonecraft euphemistically referred to that part of his business as "alum and soap."

There's a good chance that thirty-three-year-old Mary Wollstonecraft was a virgin when she met the charismatic, experienced Gilbert Imlay. Whatever the differences in their starting lines, once the pistol fired, Wollstonecraft did her best to catch up. This dawning sexuality would have been a conundrum for anyone familiar with *A Vindication of the Rights of Woman*, published just the year before. According to *Vindication*, ideal marriages were supposed to be nearly devoid of sex. In its place, Wollstonecraft argued for a relationship of mutual respect, in which "the heart, rather than the senses, is moved." In *Vindication*, she approves of "affection," but argues that "the personal intercourse of appetite . . . is despicable."

During the more chaste periods of her life, Wollstonecraft looked at sexual passion as an elaborately decorated outhouse. Sure, what goes on behind that fancy door is necessary, but for God's sake, let's not dwell on it. Then she *had* sex and was like, *Come again?* (Sorry, that was too easy.)

"After years of preaching deferred gratification and rational living, sexless marriage and pure affection, Wollstonecraft had complemented the French Revolution with her own revolution: she had entered a sexual relationship outside marriage." And she hadn't just "entered a sexual relationship"; she'd cannonballed into the pool, utterly jettisoning the sexual ideals from *Vindication*. Now sex wasn't a disease infecting a marriage of mutual affection. It was a basic human experience, "like hunger or thirst." *Exactly*, say Second Wave feminists in the 1960s and '70s. *Gasp*, say readers in 1798.

Wollstonecraft even went so far as to change the meaning of a fairly important word: "[She] now saw sexual desire as natural and right for women; indeed she went on to affirm that 'chastity consisted in fidelity and that unchastity was an association with two people at the same time.'" That is a moral shift of tectonic proportions. Wollstonecraft was saying that premarital sex is totally fine and that things go morally askew only when you're involved in more than one sexual relationship at a time. Wollstonecraft's open-mindedness would have been out of place even two hundred years later, in a country that had undergone a sexual revolution, let alone in eighteenth-century Britain. For that reason, she stayed in France with Imlay, rather than returning home. France didn't have time to hem and haw over whom Mary Wollstonecraft was sleeping with. It had its own guillotine-y problems.

There was another reason Wollstonecraft was reluctant to return to England. In May 1794, she gave birth to a girl named Fanny Imlay. Because Robespierre and the Terror were in full swing in Paris, and British citizens were being arrested and jailed, Wollstonecraft had taken Imlay's last name as a form of protection. To her, this implied a marriage-level commitment. Imlay, however, saw things differently. Within weeks of Fanny's birth, he departed revolutionary France for London. Wollstonecraft was left behind.

Thus begins one of the more tragic episodes that *Memoirs* lays out before the public in excruciating detail. "She had expected his return from week to week, and from month to month, but a succession of business still continued to detain him." Time and again, Imlay promised to return to Wollstonecraft. Time and again, he did not. After a particularly harsh and dangerous winter in France alone with her child, Wollstonecraft returned to London in 1795, only to find that Imlay had moved on to other paramours.

Wollstonecraft left a note and drank poison (presumably laudanum). Whether before or just after she'd imbibed, Imlay intervened, and the suicide was averted. Five months later, she finally realized that any reconciliation with Imlay would be impossible, and she made the second attempt on her life by throwing herself from Putney Bridge into the River Thames.

Far from William Godwin's original intentions when publishing *Memoirs*, the public's reaction of moral horror to Wollstonecraft's sexual promiscuity destroyed any chance of a fair reading. Godwin, himself a controversial and well-known philosopher, may have tried to turn things around at the end by describing his generally happy courtship and marriage to Wollstonecraft. But every ceremonial sword he unsheathed was double-edged. He let slip (or rather came right out and said) that he and Wollstonecraft shared a passionate premarital sexual relationship, that said relationship resulted in a child, and that he married her after the pregnancy so she wouldn't be ashamed in front of London society. It didn't take long for critics to pounce on these scandals.

While the shiny apple of economic independence hung before professional authors, women such as Wollstonecraft who managed to get their hands on such fruit found theirs laced with poison. Even the most celebrated female authors were respected according to their ability to "value female modesty and morality above literary ambition." According to literary scholar Catherine Ingrassia, "The eighteenth century began its own process of writing a kind of women's literary history. Publications [favored female] writers who exhibited propriety, modesty, and decorum." On the flip side, if you're a woman and you scorn society's values, then your contributions are blasted to pieces through morally inflected attacks. Thus Wollstonecraft, author of one of the greatest works on women's rights, is reduced to the following verse about the "dame the Rights of Women writ":

Lucky the maid that on her volume pores,
A scripture, archly fram'd, for propagating w---s.

The periodical that published this verse disliked Wollstonecraft for her liberal politics, and it wasn't above demonstrating its disdain by attacking her sexuality. In its debut issue, the word *Prostitution* was cross-referenced to "*See* Mary Wollstonecraft." Once the famous female rationalist, now she was reduced to nothing but a sexual punch line.

The extreme humanizing of the Mother of Feminism was just too much for, well, everyone, for the next hundred years. Never mind that *Vindication* laid the groundwork for women's rights all over the Western world. Wollstonecraft had not one, but two premarital sexual partners and periodically suffered from depression. In an era when female authors were often written off as emotional "scribblers," the passions of Mary Wollstonecraft were just too easy a target.

Wollstonecraft was not just forgotten. She became *persona non grata*. If people spoke of her at all, it was as a cautionary tale. Considering her traumatic death, that tale would probably have been a German fairy tale, and all the children of Europe would have wept and promised never to give women equal rights because look how it turned out for Mary "the Bridge Jumper" Wollstonecraft.

Mary Hays, another British feminist and an acquaintance of Wollstonecraft and Godwin, left Wollstonecraft's name out of a six-volume history of more than three hundred famous women published in 1803. As one scholar points out, "Hays's actions here remind us of just how dangerous it had become by 1800 for a woman who hoped to be published and taken seriously to identify openly with Wollstonecraft *as a person*." Because a woman's standing could not survive a scandal of the magnitude of Godwin's *Memoirs*, other women's rights advocates, "who did not wish to be tarred and feathered with the blackened brush of Wollstonecraft's reputation," felt they had no choice but to excise Wollstonecraft from their cause.

In this climate, Wollstonecraft's due as an author was initially revived not through her own works, but through later biographies. Ironically enough, her rehabilitation started with someone who initially intended to write a biography of Godwin. Charles Kegan Paul's 1876 biography of William Godwin focused on many overlooked details of Wollstonecraft's life, but whitewashed much of her sexual history. These well-intentioned lies achieved what Godwin's well-intentioned truths could not.

New editions of *Vindication* began to appear from the 1890s onward. With each new edition, commentators have continually resurrected and reinvented Wollstonecraft to suit the needs of their generation. Even our current scorn for Wollstonecraft's sexist treatment is more a reflection of our society than hers. Her reputation will continue to change as long as our society and our concept of authorship do the same—which is to say, forever.

Long before the invention of the Western printing press, the definition of "author" had been in flux. Maybe the author was a prophet, standing in the place of the gods. Or perhaps he was simply an entertainer, imitating for a living. An author might be a vessel, merely recording the inspiration of God as it flowed through him. Or he might be a professional writer, selling the products of his mind for cash. Near the time of Wollstonecraft's death, the author became an individual genius, standing far above his fellow human beings.

Mary Wollstonecraft did her best to change all the *he*s in that last paragraph to *she*s. As one scholar aptly puts it, "For women who had no rights, no individual existence or identity, the very act of writing—particularly for a public audience—was in essence an assertion of individuality and autonomy, and often an act of defiance. To write was to be; it was to create and to exist."

At a time when the world recoiled at the complicated emotions of human relationships, women such as Mary Wollstonecraft lost their words, their identity, and even their historical existence. Because the words *author* and *authority* have always had a tenuous relationship, it would take another century to begin to restore Wollstonecraft's feminist nom de guerre.

Wollstonecraft's writings mean something different to each person and each generation. This means that in many ways the story of Mary Wollstonecraft is really the story of ourselves. Charlotte Whitton, the first female mayor of a major Canadian city, once remarked with characteristic acerbity that, "Whatever women do[,] they must do twice as well as men to be thought half as good." In 1798, however, Godwin's biographical gamble turned out to be a poor one indeed. Whatever else he may have predicted about his wife's reputation when he published *Memoirs of the Author of A Vindication of the Rights of Woman*, he didn't understand that a prophet cannot pursue gratification. A vessel cannot be flawed. A professional cannot be desperate. And a genius cannot be weak.

Well . . . if you're a woman, anyway.

CHAPTER NINE

9

AMERICAN BOOKANEERS

ONE HUNDRED TWENTY-TWO YEARS BEFORE screaming crowds would welcome the Beatles to New York, American fans waited with similar anticipation in Boston for the arrival of the biggest cultural celebrity of their day: Charles Dickens. This visit was probably the greatest thing that had ever happened to Boston Harbor, next to the vandalism that started this country in the first place. In the very least, it had been a while since anyone in the United States was this thrilled about a Brit setting foot on American soil. Just twenty-eight years before, in 1814, England had invaded Washington, DC, and set fire to the White House. But this was Charles "Defender of the Downtrodden" Dickens, baby. He may have called London home, but he was one of us. He wrote stories that spoke to our lives and situations, books that transcended state and economic borders. Dickens was "a phenomenon, an exception, a special production." He was Beatlemania more than a century before the Beatles. Unfortunately, he also lived in an era before international copyright. So while he was enjoying Paul McCartney's fame, he certainly wasn't enjoying Paul McCartney's fortune.

The American press loved Dickens. While a little grumpy about the throngs of worshippers in the streets, the *New-York Evening Post* seemed genuinely proud that "a young man, without birth, wealth, title, or sword, whose only claims to distinction are his intellect and heart, is received with a feeling that was formerly rendered only to emperors and kings." The *Morning Courier and New-York Enquirer* ran a story just after he arrived, noting that Dickens was receiving attention beyond many "distinguished persons who had never been so feted by the nation, among them the 'illustrious *WASHINGTON*, the Father of this Country,' and '*LA FAYETTE* . . . a public benefactor, and the Nation['s] guest.'" In the nineteenth century, the world of print had exploded into the mainstream culture of the English-speaking world. And Dickens was at the center of it all.

Charles Dickens didn't inherit his noble status. He earned it with his own labor and unique talent. He was the embodiment of the American Dream. He was British, sure—whatever—but so were most American Dreamers just a few decades before. "His mind is American," boasted the *New York Herald*, "his soul is republican—his heart is democratic." What greater act of narcissistic affection can there be than to shout as a nation, *We love you, Mr. Dickens! We love you because you are we!*

For his part, Dickens returned that endearment. Upon settling into his room at the opulent Tremont House, he descended the staircase with his wife, jauntily announcing to his adorers, "Here we are!" Writing later, Dickens declared, "Boston is what I would have the whole United States to be."

That's a nice sentiment, Mr. Dickens, but here's what Americans would be saying about you in just a few weeks' time: "cockney," "literary bagman," "penny-a-liner loafer." In case the outdated insults are a bit too obtuse for a modern audience, allow us to continue: "[Dickens is] the most flimsy—the most childish—the most trashy—[and] the most contemptible . . . [his writing is the] essence of balderdash reduced to the last drop of silliness and inanity."

And then there's this: "the sooner [Dickens] hangs up his fiddle, and himself with it, or jumps into the New River, the better it will be . . ."

Ouch. What the hell happened? Dickens's sudden fall from American grace was reminiscent of the Beatles' fall in 1966, when Lennon declared to a British daily, "We're more popular than Jesus." That went over in the American South about as well as anyone could expect. Dickens avoided making any comments about the god of Christianity, but he did say a few words about the god of America: Cold. Hard. Cash.

LIKE AUTHORSHIP, the idea of copyright as we understand it today has experienced an ongoing evolution. In 1842, America had no international copyright agreements. Copyright protection did exist (after the pattern established in England in the eighteenth century), but it was restricted to American citizens. This meant that writers from other countries, even the most celebrated, had no claim to the profits of any book printed and sold by American publishers. Nor did they have any means of stopping those publishers from producing them. Of course, the lack of reciprocal international copyright laws wasn't doing any favors for American writers, either. It was to this more convincing point that Dickens first broached the subject of copyright, at the end of a speech in Boston about a week after he arrived. Initially, his comments were received with a flood of applause from the attendees. But within a few days, that sentiment had taken a decidedly worse turn.

With his first major hit in 1836, Charles Dickens achieved success right when literary celebrity was becoming popular. Dickens scholar Robert L. Patten explains that the author "thought of himself as a professional writer, an identity which an older generation deplored." As scholar Florian Schweizer notes, "Due to the success of a few writers, a whole generation of readers started to believe that literature was the key to fame and fortune. Writing in 1837, the journalist James Grant confirmed that authorship had become a commonplace occupation: "'You now meet with an author in every fifteenth or twentieth person you chance to encounter in the daily intercourse of life.'"

With the blossoming popularity of authors in the Romantic Age such as Lord Byron and Sir Walter Scott, publishers began to capitalize on the idea that customers would buy books simply because of a name. This was an important shift. An author's name was part of the overall sale package, one more tool to move the product. Sooner or later, authors were bound to ask for a fair cut of that product.

In the first centuries of print, very few living authors wielded power by their name alone. Martin Luther was one; Shakespeare was another. Yet, as popular opinions of authorship changed, writers were able to use the expanding economy of print to seize more monetary rewards. This trend was part of a wider Enlightenment ideal that praised the right to property rather than regarding wealth as a source of depravity. Even ideas, by their very nature intangible, were slowly being accepted as property. That McDonald's could own the word combination "I'm lovin' it" would have been a revolutionary concept before the Enlightenment.

John Milton, for example, sold his manuscript of *Paradise Lost*, and his right to profit from any proceeds of the work, for five pounds in 1667. No matter how unfair this seems today, at the time it was business as usual. Authors produced and sold stories just as a farmer might raise a cow and sell the beef to McDonald's, which then "processes" it for sale to a larger group of consumers at a profit. Because

English protocopyright structures were built around literature as a single act of creation, laws protecting the rights to literature were often meant to protect those in charge of distribution, essentially the publishers of the day, not the original creator of the work.

As England's philosophical and economic underpinnings shifted with the Enlightenment, so, too, did its view of copyright. A major breakthrough occurred in 1710. This was the year of the Statute of Anne, an English law entitled "An Act for the Encouragement of Learning, by Vesting the Copies of Printed Books in the Authors or Purchasers of such Copies, during the Times therein mentioned" (hence why everyone agreed to call it the Statute of Anne). This act granted copyright holders rights to a work for a period of fourteen years. The protection could be extended one time, for another fourteen years, upon request. Boom. Problem solved. Before 1710 an author had legal recourse only in cases of "libel, blasphemy, or sedition" (where he was usually the criminal defendant, by the way). After the Statute of Anne, the author was now a legal entity who had ownership of his works. The full implications of this took a little time to sink in, but moving forward, reproductions could be printed only by permission of the author, or by someone to whom the author had sold the "rights" to his personal "copy." Violate that, and an author finally had grounds to sue.

Foreigners were another story, however, and in the nineteenth century, that story began to change. International agreements were emerging among various developing nations, but one country that bucked that trend was the good ole U.S. of A. "Inasmuch as [Dickens] received 40,000 guineas per annum for his works in England . . . he had no right to ask Americans 'to double the sum out of [their] own pockets.'" This was Boston's response, in *American Traveller*, to Dickens's calls for copyright reform.

First off, 40,000 guineas seems a bit high, even for Mr. Dickens. This staggering annual income would be roughly equivalent to £1.7

million in modern currency. Thinking Dickens had access to that extreme amount of capital was mistake number one. Yet, with that error firmly planted in their minds, it became pretty easy for people to slam the door in his face when he came asking for money. It would seem a little like Ebenezer Scrooge holding out Oliver Twist's bowl and saying, "Please, sir, I want some more—money, that is."

Publishers in America who were pirating Dickens's books shared that indignation. Why should Dickens get paid twice? He'd already gotten his guineas back in England. Why should we Americans give him money in exchange for products that we consume? Sure, that's the basic definition of capitalism, but you can just bugger off, capitalism, Dickens has enough money.

In the 1840s, America was the literary equivalent of the *Queen Anne's Revenge*, and the printers, publishers, and booksellers of this country were the ship's captain, Edward "Blackbeard" Thatch. We were pirates, and what's worse, we were proud of it—really, really proud of it. To deprive foreign authors of royalties was the American way. We were so famous for it that a new phrase was coined in the history of print just for us.

"No sooner is a literary venture of Bulwer[-Lytton], Thackeray, or Dickens afloat, than a whole baracoon of 'bookaneers,' as Hood called them, rushes forth to seize it."

Bookaneers. Oh my, that's clever. It makes you think of a whole fleet of well-read pirates mounting raids on libraries to pore over the treasured works of Jonathan Swift, Goethe, and William Wordsworth. And, in fact, American publishers liked to make similar claims as their motivation for piracy. See, books are cheaper when they're stolen. Publishers claimed that pirating works allowed their prices to remain low, which in turn made the works more accessible to the public at large. So piracy = educating the masses. It's an Enlightenment ideal. American printers were proud to be pirates because they pictured themselves as taking part in the education of the infant nation, just as they had taken part in its infant democracy.

Whether Dickens went to America with the express purpose of advocating for international copyright has been a matter of intense speculation and debate from the moment he arrived in Boston. Whatever his intentions may have been, he appears to have stumbled naïvely into a very awkward situation with the United States. It was like a first date. Both people have lofty expectations and dress to look as sharp as possible. They smile and say all the right things, but after a while they begin to realize that something's off. It becomes clear that one of you isn't as wealthy as you said you were, and things get really weird when he starts asking you for money. Then you get offended when he visits Richmond, Virginia, and hates all your slave plantations. Classic first-date hiccups.

Awkward or not, Dickens's 1842 visit shone a light on the problem of American copyright. As straightforward as the solutions seem today, pirates controlled the vast majority of the printing presses in nineteenth-century America. As a result, literary boarding planks became symbols of patriotism.

I HAVE MADE UP MY MIND . . . TO GO TO AMERICA," Charles Dickens cap-shouted to a friend in a letter dated September 1841.

Why did he want to go? Who knows? He never told anyone. A few historians have claimed that, from the very beginning, he planned to proselytize for copyright reform. American newspapers would claim that Dickens was a "mercenary," a hired thug sent out by a cabal of British authors from some Literary Legion of Doom, which, let's be honest, is the nerdiest of doom legions. Maybe he just wanted to pay a visit to his pen pal Washington Irving, or find out what a hushpuppy was, or see the PIGS on Broadway. (Sorry, that made it seem like there was a show called *The Pigs* on Broadway in 1841. We were just caps-shouting about the roaming passels of hogs that regularly foraged along Broadway—something Dickens found amusing enough to write home about.)

Whatever initially carried Dickens to the United States, one thing is for certain, he got here in style, on a fancy steamer, christened the SS *Britannia*, that could make the transatlantic crossing in just fourteen days. This leads us to gaffe number one on Dickens's first date with America. We thought he had money, but he did not.

Because Dickens was "the most photographically famous person in Britain outside of the royal family," everyone knew what he looked like. "[Dickens] was instantly recognizable on the streets, furiously walking, in flamboyant clothes." If you closed your eyes and pictured Charles Dickens, this is what you saw: "a middle-sized person, in a brown frock coat, a red figured vest . . . a fancy scarf cravat, . . . [a] dickey . . . a double pin and chain . . . a gold watch . . . a shaggy coat of bear or buffalo skin that would excite the admiration of a Kentucky huntsman."

His fashion style has been described as "graphic fancy." On one leg of his American journey, he was actually mistaken for a riverboat gambler. That's right, in comparison to the people in his immediate vicinity, Dickens looked like some fancy Dan trying to stroll off the pages of a Mark Twain novel.

When Dickens traveled, he did so in style. He boarded in the stateroom of the SS *Britannia*. Once on land, he stayed in presidential accommodations such as the Tremont House (America's "pioneer first-class hotel") and the Clifton House Hotel (where his "bedroom and sitting room windows looked straight down upon [Niagara] Falls").

Wherever Dickens went he exuded fame and fortune. Still, fortune is not the inevitable successor to fame. Just ask any flash-in-the-pan YouTube celebrity. In reality, Charles Dickens was broke.

Worse than broke, he was £5,019 in debt. The only way he was even able to raise the capital for the American excursion in the first place was to take a loan out on his next novel. On New Year's Day 1842, he acknowledged receipt from his publishers of £885, and three weeks later pushed off for Boston.

When Dickens and his English publishers issued what turned out to be his breakout success, *The Pickwick Papers*, in serial parts (1836–37), they sold for a shilling a piece. (A complete set in parts now sells for thousands of dollars, but that's beside the point, especially for Dickens.) This method of selling and distribution created a huge potential for profit, "transform[ing] Dickens, Chapman and Hall [his publishers] from minor figures in Victorian letters to titans." Yet the author-publisher relationship as we know it today was still maturing, and Dickens saw exceptionally less personal profit from his early successes than a modern-day author would. Even with his unprecedented sales numbers, he wasn't financially secure until 1847. That's ten years after *Oliver Twist* (now worth tens of thousands of dollars), and three years after *A Christmas Carol* (now worth as much as thirty-five thousand). When Dickens, arguably the most famous author in the Western world, landed in Boston in 1842, financial security was still half a decade away.

Despite writing five runaway hits and almost single-handedly revolutionizing the way books were sold, Dickens had only ever received about three hundred fifty pounds from American publishers. Compare that to the quite modest two thousand pounds he made in England from *The Pickwick Papers*. With IOUs bursting out of his buffalo-skin coat pockets, it's not hard to see how he might have felt compelled at least to mention copyright reform when a soapbox was presented to him. If Americans really were being expected to "double the sum" of forty thousand guineas annually "from their own pockets," as the *American Traveller* complained, we were kind of admitting to fleecing Mr. Dickens out of two hundred thousand dollars a year in lost copyright revenues—and that sum of money was more than enough to wage a small war over.

"Before I sit down," Dickens said at the tail end of a banquet thrown by the Young Men of Boston on February 1, "there is one topic on which I am desirous to lay particular stress." His speech up to this point had focused on things like Little Nell, the thirteen-year-old protagonist of one of his latest novels, *The Old Curiosity Shop*. A pure and gentle soul, Nell (nineteenth-century spoiler alert) doesn't fare too well by the end of her story. Dickens wasn't quite at the level of George "Serial Killer of Everyone You Ever Loved" R. R. Martin, but Nell's passing was a bitter pill for Dickens's readers. He'd been touched by letters regarding her demise from "dwellers in log-houses among the morasses, and swamps, and densest forests, and deepest solitudes of the Far West." That's sweet. Round of applause . . . oh, wait, there's more.

"I hope the time is not distant," Dickens continued, "when [authors], in America, will receive of right some substantial profit and return in England from their labors; and when we, in England, shall receive some substantial profit and return for ours."

He then ended with "I would rather have the affectionate regard of my fellow men than I would have heaps and mines of gold. But the two things do not seem to me incompatible."

Essentially he was saying, *Hey, guys, all I need is your love and support. I don't actually need gold chains and bear coats and silk scarves and dickeys, but . . . I mean . . . there's no reason I can't have both, right?*

Dickens closed to "tumultuous" applause from the audience. Maybe the tumult was for Little Nell, or maybe the Young Men of Boston really loved speeches about intellectual property rights. Either way, Dickens brought up an important point, one that he believed would resonate with American audiences: just as he was receiving almost nothing from the American sales of his novels, without a strong international copyright agreement, American authors were also being shortchanged abroad.

That should do it, right? If you want to rouse Americans, kick them where it counts—right in the coin purse.

He clearly hadn't spent enough time with Americans. As one scholar summarizes, "[Dickens] mistook piracy for the result of administrative oversight, which could easily be rectified by a single piece of legislation." This is probably how he pictured things going in his head:

Pardon me, people of America, it appears you have forgotten to pay me.

We'll just look into that for you, Mr. Dickens. Good Lord, man, you're right! Please accept our humblest apologies. And also this mountain of dollars.

Here's how it actually went, in the New York's *Morning Post*: "[It was] bad taste in Mr. Dickens to allude to the copyright business in his speech here."

Such "pecuniary considerations" . . . "smells of the shop—rank," railed the *Boston Morning Post* and the *Courier and Enquirer.*

Dickens was uncouth to "intrude his *business* upon those who assemble to pay homage to his genius," tsk'd the New York *Courier.*

"[W]e want no advice on this matter, and it will be better for Mr. Dickens, if he refrains from introducing the matter hereafter," threatened the *Hartford Times.*

Dickens either didn't get those threats or didn't care, because a week after his Boston speech, he brought the subject up again at a public dinner in Connecticut, with his most winning Victorian graces: "I would beg leave to whisper in your ear two words, International Copyright."

Seeing as the economic tack hadn't worked, Dickens retreated to his specialty: emotional appeal. What if withholding royalties from an author were enough to kill him? Could American stinginess murder an author as thoroughly as the cruelties of life murdered Little Nell?

Dickens's predecessor, the revered novelist, playwright, and poet Sir Walter Scott, had died near penniless in 1832, never receiving so much as "one grateful dollar" from his American fans. He perished tragically, broke and in debt, without enough money "to buy a garland for his grave." (Today, first editions of his best-loved work, *Ivanhoe*, sell in the healthy mid-thousands of dollars.) If Americans hadn't been so greedy, Dickens not so gently alluded, Scott "might not have sunk beneath the mighty pressure on his brain." In other words, America, you killed Walter Scott.

Calling us moneygrubbers is one thing, but blaming us for the death of Sir Walter Scott? Aw, hell no.

"Mr. DICKENS has been honored with two public dinners since his arrival in the United States; and on both occasions he has made an appeal to . . . *dollars and cents* for his writings. We are . . . mortified and grieved that he should have been guilty of such a great indelicacy and gross impropriety . . . by urging upon those assembled to do honor to his *genius*, to look after his *purse* also!"

Indelicacy! Impropriety! Mortification and bereavement! It's bizarre to think that a country built on capitalism would feel so importuned by Dickens's request to be compensated for his work in "dollars and cents," rather than hugs and kisses and unicorn smiles. But the state of copyright in the United States was a quagmire whose depths Dickens seemed genuinely ignorant of. Four days after his Hartford speech, the *New World* commented with unusual insight that "time, place, and occasion taken into consideration, [Dickens's copyright remarks] seem to have been made in the worst taste possible." Some of that American pushback can be explained by delicate events unfolding between the United States and Britain at the time. The rest can be explained by our printing presses. In particular, our publishers, who had "indelicate" aspirations of their own.

At the time of Dickens's visit, America was at the tail end of a massive recession. Around 1834 the United States entered a boom economy, fueled largely by rising prices for land, cotton, and slaves. As bubbles are wont to do, this one burst, around the summer of 1836. With banks locking down on loans, and money shifting away from commercial centers in the East, America smacked into the Panic of 1837. Then "in 1839 panic set in again: a depression whose severity, according to one recent economic historian, 'can most accurately be compared to that of 1929.'" The U.S. printing industry was hit particularly hard during these times.

Historian James Barnes sets the scene: "Perhaps nothing in the nineteenth century so influenced the American book trade as the depression of 1837–1843 . . . New publishers sprang up only to disappear a few years later amidst the ranks of debtors and insolvents. Editors moved from one journal to another, seeking to stave off the inevitable. Prices for books and periodicals fell lower and lower."

Consider that, in the 1820s, books were priced in the two-dollar range. By the late 1830s those same books were selling for fifty cents—a staggering three-quarter price drop. How could the American publishing industry survive on such meager rations? Well, it did what most people are willing to do when they're starving: steal.

In the midst of this economic recession, publishers came up with the idea of printing books in newspaper form. "Weeklies" were huge folio sheets of printing paper that could "go through the post at newspaper rates instead of those for magazines [or books], and such large sheets could be cheaply printed, requiring no binding or stitching, and yet had more reading matter than most literary periodicals."

What kinds of materials could these publishers print cheaply in their weeklies? The stolen kinds, obviously. *Brother Jonathan*, based in Manhattan, was one of the first and largest of these weekly publications, and proudly boasted of "having first introduced into the cash newspapers the custom of reprinting [Dickens's] novel[s] as they appeared in numbers." These weeklies became a national phenomenon. Life in the publishing industry suddenly started looking up. U.S. printers had access to a constant stream of novels, paid no royalties to the authors, and delivered those novels to American readers in the absolute cheapest way possible.

As far as publishers in the United States were concerned, wholesale pirated material was a win-win for everyone involved. Publishing houses started seeing profits again, American readers were given access to novels they couldn't have afforded otherwise, and foreign authors were granted the greatest gift of all: love. "The community . . . owes us a debt of gratitude," crowed the *New World*.

Let's be clear: Americans were the worst copyright offenders in the Western world. The theft of intellectual property that started out as a means to "jumpstart a new, more enlightened and democratic polity" had quickly become a pirated-version of Frankenstein's monster—a Maggie Shelley's *Frankenstone*, if you will. Dickens's works were pirated so ubiquitously that you could find them on the back of railway timetables. This was an industry supported by more than two hundred thousand jobs, printing off three to four times as many books as Britain. Granting copyright protections to foreign authors would have been tantamount to economic *hara-kiri*. The very notion was patently absurd to U.S. publishers.

There is some truth to the claim that Dickens's popularity in the United States was greatly fanned by the cheap availability of his novels. If he continued asking for those "dollars and cents," U.S. printers threatened, "it was to be subtracted from the earnings of publishers whose presses had purportedly made the writer famous enough to be received as America's guest."

Yet the patriotism of those printers lasted only as long as it increased their profits. Homegrown American authors were hurt by these policies as well. Because domestic copyright had already been established, American authors had to be paid for their work (*gasp*). But why would a publisher bother with an inconvenience like that when he could just live off the fat of English bestsellers? Pirated works were so cheap that they were sold, according to Mark Twain, "at prices which make a package of water closet paper seem an 'edition de luxe' in comparison." Books by American authors generally cost four times as much as pirated books. It's as if books by English authors were being sold at used paperback prices, while American authors were sold exclusively at hardcover.

A growing number of American authors added their voices to the fight for copyright. For instance, the American poet James Russell Lowell attempted to cast shame on cutthroat publishing practices with his matronly slogan, "Better than a cheap book, is a book honestly come by." *Ha ha ha, shut up, Lowell. Sincerely, the American printing industry.*

Now, if you're thinking there should be more xenophobia in these nineteenth-century discussions, never fear: "The country is drugged from one end to the other with foreign literature which pays no tax," complained Washington Irving. According to the American Copyright Club, British culture "sweeps the land, and puts at naught all petty distinctions of district and neighborhood, and settles down, at its leisure, into a dark, slimy, universal pond." Competing with this flood of cheap, abundant, British literature was even described by inventor Samuel Morse as an ongoing "colonial bondage."

Publisher Roger Sherman put it succinctly when he antagonistically wrote, "[Copyright reform is] the clamor of two hundred authors against the interests of fifty-five millions of people." It turned out that two side effects of the 1837 recession had revitalized the American publishing industry: giant, cheap weeklies and teeny, tiny violins for authors.

After making a third attempt to address copyright reform at a

New York banquet on February 18, Charles Dickens basically threw in the towel. This was the moment, during his first date with America, when he realized we were totally incompatible. "This is not the Republic I came to see," he wrote. "This is not the Republic of my imagination." Or in other words, *You look totally different from your profile pic, America. It was obviously taken eight years ago and under noticeably better lighting.*

On America's end, Dickens was quickly living up to the first four letters in his name. As his disappointment in us grew, so, too, did his criticisms—much of which came in the form of his travelogue, *American Notes for General Circulation*, published a few months after his return to England; and his next novel, *The Life and Adventures of Martin Chuzzlewit*, which began serial publication three months after that. When Dickens first conceived this picaresque novel, it was not supposed to take a detour to America. But after returning home, Dickens had a few complaints fresh in his mind. Americans were not portrayed in a particularly favorable light.

On a steamboat, "every gentleman on board appeared to have had a difference with his laundress, and to have left off washing himself in early youth," Dickens wrote. Oh, Dickens: laundresses were for people with silk cravats and dickeys.

Our favorite scene is the description of one American dinner where a bell summons feasters to the table:

> Three more gentlemen . . . came plunging wildly around the street corner; jostled each other on the steps; struggled for an instant; and rushed into the house in a confused heap of arms and legs . . . All the knives and forks were working away at a rate that was quite alarming; very few words were spoken; and everybody seemed to eat his utmost in self-defense, as if a famine were expected to set in . . . The poultry . . . disappeared as rapidly as if every bird had had the use of its wings, and had

> flown in desperation down a human throat . . . Great heaps of indigestible matter melted away as ice before the sun. It was a solemn and an awful thing to see.

And Dickens wrote that without ever setting foot in a Golden Corral, or the state of Texas. By the end of his fictional romp through the United States, Dickens's protagonist Martin Chuzzlewit concluded that most Americans lack "that instinctive good breeding which admonishes one man not to offend and disgust another."

Before arriving in America, Dickens expected to move through our streets and boroughs as a witty little observer, much as he had in London. In reality, though, he was constantly crushed by a throng of fans wherever he went. At one point, the riverboat carrying him and his wife docked in Cleveland for the night, and the next morning "a party of 'gentlemen'" competed with each other to peer inside the cabin window "while [he] was washing and Kate lay in bed."

He received hundreds of requests for locks of his hair. One New York barber who serviced him turned around and immediately capitalized on this by offering his hair clippings for sale.

"I can do nothing that I want to do, go nowhere where I want to go . . . If I turn into the street, I am followed by a multitude. If I stay at home, the house becomes, with callers, like a fair . . . I take my seat in a railroad car, and the very conductor won't leave me alone. I get out at a station, and can't drink a glass of water, without having a hundred people looking down my throat when I open my mouth to swallow."

Perhaps some of those mouth gazers were looking to see if the ads in one newspaper were true. "His 'rather yellow teeth,' showed, said the Philadelphia *Spirit of the Times* . . . that he did not avail himself of 'Teaberry Tooth Wash' or 'Hufeland's Dentifrice,' the virtues of which commodities were described elsewhere in the paper." You know Americans love you when they break out the merch. Using

Dickens's early pen name, one New York grocer advertised "Boz Pork & Beans." Nothing gets you in the mood for *Oliver Twist* like a can of gelatinous legumes and pork fat that only starving children would consider edible.

At first, Dickens embraced this deluge of American attention. Then he tolerated it. Then he hated it. After a few weeks, our incessant rudeness came to represent all that ailed American society. The "mass of your countrymen," Martin Chuzzlewit announced, "begin by stubbornly neglecting little social observances . . . acts of common, decent, natural, human politeness . . . From disregarding small obligations, [Americans] come in regular course to disregard great ones"—such as copyright reform, the plight of the poor, and the elephant in the room: slavery.

Charles Dickens was supposed to be our great English ambassador of goodwill. At the time of his visit, tensions had been mounting between the United States and Britain over issues such as the Canada-Maine boundary, the Oregon boundary dispute ("Fifty-four Forty or Fight!"), and England's search and seizure of American slave ships. (Britain had abolished slavery as an unlawful evil in 1833.) But Dickens was supposed to fix all that! He was supposed to show the world how awesome we really are! Instead, he wouldn't shut up about copyright, and then called us all gluttons bereft of human politeness. He offended our capitalism just because we were stealing, and offended our national character just because we were rude.

In a display of patience and generosity, we even invited him to tour a tobacco slave plantation in Virginia. That's right, a premium tour of a slave plantation. For the foremost literary activist regarding society's most abused classes. According to his biographer Michael Slater, "[Dickens] could hardly have failed to be struck by the incongruity of being toasted in a slave-owning community as a writer whose work makes us 'feel for the humblest' and 'creates in all of us a sympathy for each other—a participation in the interests of our common humanity, which constitutes the great bond of equality.'" In disgust, Dickens canceled all future travel plans in the American South.

During the last legs of his American journey, Dickens traveled as far into the frontier as St. Louis, Missouri, a place he found to be "intolerably conceited." On his return, he stayed at Niagara Falls, whose majesty helped to dull his disappointment in this transatlantic trek. It was at the Falls that he "at last found that sublimity of nature he had been so ardently hoping to experience in America." It was also here that he received an open letter addressed to the American people and signed by twelve of the most influential authors in Britain, as well as a separate letter by historian and satirist Thomas Carlyle. The letters strongly argued for an international copyright agreement between the two countries.

Dickens forwarded these letters to newspapers in New York, Boston, and Washington, DC, and to J. P. Kennedy, a pro-copyright congressman from Maryland. The letters had little to no effect. Some contemporaries have looked at this move through a lens of perfidy because the letters were sent to various news outlets without Dickens explicitly stating that he'd requested that they be written. What appeared to be a spontaneous and collective protest from the British Isles had in fact been orchestrated by Charles Dickens.

Perfidious or not, the letters were a feeble volley as Dickens retreated from the American press corps. By the end of his trip, the same newspapers that had welcomed him like royalty were calling him an "adventurer," a "mercenary," and a hypocrite—it bordered "on the ridiculous for [Dickens] to lecture Americans . . . about dollars, who is clearly convicted of a supreme love of them." The "sacred wrath of the newspapers," as Dickens put it, had won.

Arriving back in London on June 29, 1842, Dickens took out his revenge in the most effective way possible—by using our craving for celebrity against us. *American Notes* lambasted the United States. *Martin Chuzzlewit* takes that unplanned detour to America just to denounce our gluttonous, boorish ways. And we bought it. In droves we consumed it. As much as he railed against America, Charles Dickens was still the most popular author in the English-speaking world. *American Notes* became a best seller in the United States; so did *Martin Chuzzlewit*. No matter what Dickens put in his tales, we had no choice but to eat it. It tasted too good to stop. Like an entire box of Little Debbie Oatmeal Creme Pies, we cried and ate, and cried and ate some more.

The newspapers that crucified Dickens's character couldn't help themselves from including even his most caustic stories within their pages. To do otherwise would have been to slit their wrists and spill so many moist dollars on the floor. "[Printers in America] were driven by popular demand and their own covetousness to go on stealing the very text they would have to denounce as slanderous in other parts of their . . . newspaper." Dickens wouldn't get financial recompense for his stories, but who can put a price tag on throwing shade at your haters, and knowing they have to print it on page one?

AS THE awkward first date with America receded into painful memory, Dickens might have cast his eye across the pond to see what his copyright crusade ultimately accomplished. The answer would have been nothing.

Wait, no: nothing would have been better than what *actually* resulted from Dickens's American tour. Panicked by the torchlight being shone on their pecuniary piratical printing practices (say that five times fast), members of the book trade in Boston gathered at the Boston Museum on the night of April 26, 1842, for an emergency meeting.

These pillars of the Boston printing community met to draft an official letter to Congress petitioning against *any* species of international copyright reform. Dickens was still making his way back to New York from the western frontier at the time, and even though he'd already lost the American copyright debate, Boston wanted to mount his head on the Old North Church.

The American printers requested that any imported books face duties. And to make the situation for British authors infinitely more insulting, the Boston letter claimed that Americans must be free to adapt English works "to our own wants, our institutions, and our state of society."

Wow. Do not mess with printers. Dickens went from Americans stealing his work to having tariffs levied against his overseas book sales, to publishers *changing his stories* to better accommodate American "wants and institutions." Incidentally, that's something, according to Dickens, they'd already been doing. We're not sure what changes were made, but we imagine they went something like this:

American *Oliver Twist*—Fagin explains that his gang of pick-pocketing children is entirely necessary to jump-start the local economy.

American *A Christmas Carol*—The ghost of Marley appears to Ebenezer Scrooge and warns him to raise the minimum wage, but that it doesn't actually need to be high enough for the Cratchits to pay rent.

American *A Tale of Two Cities*—This becomes a story about who has better pizza, New York or Chicago. Since it's a work of fiction, Chicago wins.

For the next fifty years, American printers were lawfully able to plunder the literary treasures of the world. Voices were raised from time to time, but economic incentives for piracy were just too powerful.

In 1886, representatives from ten countries signed an international treaty at the Berne Convention, which granted automatic copyright protection to authors regardless of their country of origin. The United States sent observers to Berne but refused to become a signatory. Five years later, in 1891 (twenty-one years after Charles Dickens died), the U.S. Congress passed the Chace Act, which finally extended copyright to select countries (including Britain) as long as their work was "simultaneously published or 'manufactured' in the U.S." America did eventually join the Berne Convention and extend copyright protections to everyone, but not until March 1, 1989. Yes, you read that right. From two years before the publication of *A Christmas Carol* to the authors of this book attending elementary school—that's how long it took to pass comprehensive international copyright laws in America.

If Dickens hadn't already been depressed enough about the state of American piracy in the nineteenth century, he'd probably be more depressed if we could tell him that there are many "Americas" in the world today. China and India are basically us from 1842, when it comes to stolen books, movies, music, and video games. Dickens might even enjoy the irony that American cultural products are currently pirated at a higher rate than those of any other country in the world.

As time marches on, many of the issues of copyright have become murkier, not clearer. Which brings us to the Internet. Oh, Mr. Dickens, let us tell you about the Internet. Content piracy may be the keystone keeping that whole damn infrastructure together. "We can note that today's struggles," writes one copyright historian, "are fought in terms that would be eminently comprehensible to those nineteenth-century reformers who battled over how broadly to extend rights and powers to authors."

On second thought, let's not tell Mr. Dickens about the Internet.

CHAPTER TEN

10

WHEN DOVES CRY

MOST FONT ENTHUSIASTS SHUDDER WITH horror at the thought of a certain old man strolling across the River Thames in the autumn of 1916. The dark figure labors to make his way to the railing of Hammersmith Bridge by night. His deliberate gait signals that he's carrying a heavy burden. And from his journals, we know that he's been suffering from severe depression. "I have been in the depths of despondency," the seventy-six-year-old man writes, "I have no longer, as it seems to me, any essential place or function in the world. I have played, or am playing, my last cards." That man is one of the key players of the Private Press Movement, and he is about to do something terrible.

A witness like you might try to call out to the old man, but the septuagenarian will not listen. He struggles to pull himself up onto the railing and, with a final heave, lets his burdens go. "No!" you cry out, just before the splash—and then another splash, and another and another. Your cry trails off as several splashes dot the surface of the dark river. Now the old man doesn't appear depressed at all. He snaps closed the box he was carrying and scampers off, giggling like a schoolboy.

What in the name of Johann Gensfleisch zum Gutenberg just happened?

Was the old man just littering? In reality, he was guilty of far worse than a petty dumping violation. He has just effected the destruction of one of the world's most beautiful fonts: the Doves Type. The shadowy figure behind this crime was none other than its devoted owner, T. J. Cobden-Sanderson.

Today we largely take typefaces for granted. In fact, we're kind of supposed to. Many designers argue that a font must do its "work and disappear." Yet that's not entirely accurate. We have countless typefaces, each telling its own story, each making us react differently, whether we realize it or not. Look, we've just changed our typeface to Courier New. Now our chapter looks like we're pitching a screenplay. Here's Arial. Now we've revealed that we're composing this chapter in Microsoft Word, where Arial is offered as an obvious rip-off of the wildly popular Helvetica. Franklin Gothic is considered the most recognizable, truly American typeface, so it's no surprise to see it appear on all-American movie posters such as that for *Rocky*. And Comic Sans . . . poor Comic Sans. It has a hard time being taken seriously because of its roots in comic book lettering—a genre whose merits have, until recently, been pretentiously ignored. The typographer Erik Spiekermann compares typefaces to footwear: Sure, they're all accomplishing the same general purpose, but steel-toe boots send a very different message from ballet shoes. Change the typeface and you change the message.

Some typefaces imply honesty, like Gotham. This typeface is known for its "unremarkability and inoffensability," which the 2008 Obama campaign used for its action words such as *HOPE* and *CHANGE*. This was "a type consciously chosen to suggest forward thinking without frightening the horses." Some typefaces have a trustworthy and authoritative air, such as Baskerville, which a recent study has indicated makes you 1.5 percent more likely to agree with a given statement.

Today, dozens of typefaces are available on our word processors, and hundreds upon hundreds more are easily accessible for download. Until recently, though, the ultimate font of cold beauty, the Doves Type, was probably the most elusive of all. Its owner, Cobden-Sanderson, guarded it with religious zeal. Then, rather than let his type fall into commercial hands, he destroyed it. All of it.

What led Cobden-Sanderson to hurl his legacy from a bridge is one of the more surprising stories in the history of print. As the adage goes, *If you love something, drown it in the Thames. If it comes back to you, it won't, because you drowned it in the Thames*. To really understand the reasons behind Cobden-Sanderson's destruction of the Doves Type, we have to look at his decade-long feud with ex-partner Emery Walker, and the origins of the Private Press Movement.

BORN THOMAS Sanderson in 1840 in Northumberland, England, he spent the first forty years of his life bouncing from locations as diverse as shipyards to Cambridge University, trying his hand at myriad occupations and failing because of dissatisfaction and recurring mental breakdowns. The artistic temperament in him seemed wholly incapable of dealing with the ho-hum life of a typical Victorian Londoner.

During a trip to Italy to recuperate from his latest downturn, he met Annie Cobden. In 1882 they married, taking each other's last names. Annie seemed a good counterpoint to T.J., and she encouraged him to follow his dreams. This is the history of print, so his dream was, obviously, to be a bookbinder.

To Cobden-Sanderson, bookbinding was "something which should give me means to live upon simply and in independence, and at the same time something beautiful." There are certainly worse ways to make a living. After a decade spent mastering bookbinding, the Doves Bindery opened in March 1893, when Cobden-Sanderson was a strapping fifty-three years old. Some nineteenth-century Londoners hit their midlife crisis and run out to buy a two-horsepower convertible Victoria with drop-down front benches. Others start a bookbindery. To each his own.

Cobden-Sanderson named his bindery after the building's location, a pleasant lot called Doves Place. Modern collectors buy certain books for their bindings alone, and Cobden-Sanderson's are some of the most highly sought after. They are exceptional for their aesthetic unity, the quality of their materials, and their high standard of finishing. Cobden-Sanderson's style is so valued that you can even find known counterfeit Doves bindings floating around the market, commanding high prices themselves, forgery notwithstanding.

Before about 1800, the components used for printing books remained surprisingly stable for three hundred years or more. Over the centuries, the book very slowly transformed into a popular commodity, rather than the expensive luxury of its earliest days. But by Cobden-Sanderson's day, the technology of print had experienced an unprecedented rate of evolution.

As economies and audiences shifted, consumer demand took off. Publishers needed to put out more product, and do it as fast and as cheaply as possible—which sounds like a job for your friendly neighborhood Fourdrinier paper making machine, or Koenig rotary press, or the game-changing Monotype. Every aspect of bookmaking was tinkered with to improve efficiency. The traditional world of book production—handmade papermaking, hand presses, and hand-set type—was vanishing behind a veil of steam, steel, and grinding gears.

The aforementioned Fourdrinier machine used a moving belt and felt rollers to make paper of any width or length. Thanks to this contraption, by about 1810 machine-made paper was being regularly produced and sold. Hot on its heels, on November 28, 1814, *The Times* in London printed its newspaper using newly mechanized presses developed by Frederick Koenig. Instead of the up-and-down, stamplike movement of a traditional press, Koenig placed the sheet around one cylinder, which was then rolled against a second cylinder. This double-cylinder, steam-powered press represented the first major change in the fundamentals of the printing press since its invention—and more than tripled its productivity. Now you've created a mechanized printing assembly line. These presses worked so quickly that daily newspapers were no longer constrained to eight pages per issue, the largest practical capacity that older metal types could handle.

Over the course of the nineteenth century, casting type jumped from around four thousand sorts (individual pieces of type) in a day by traditional hand molds to six thousand *an hour* on specialized typecasting machines. With the invention of the Linotype in 1884, type was cast with the press of a button, using a "hot metal" process. With this machine, printers could instantly create a "line o' type" using a two-hundred-pound appliance that was "kind of a cross between a casting machine, a typewriter, [and] a vending machine." Its companion, Monotype (casting single letters with the hot metal process), would soon revolutionlize the industry of typefaces.

These new machines, among many others, dramatically accelerated the capacity for mass production in the nineteenth century. If you're looking to create a large number of books, these machines were perfect. There are drawbacks when it comes to mass production, however. A product can be fast, cheap, or good. It can be a combination of two, but it can't be all three. In the nineteenth century, most publishers chose fast and cheap. The results were not what one would normally call "good."

Enter the rebels of the printing world. As early as 1639 there is documentary evidence of printers who worked for pleasure only—for the beauty of their craft—not for the golden god of profit. These rebels "form a distinct undercurrent" running throughout the history of print. Yet, in the Industrial Age, that curious dissent transformed into a full-fledged rebellion. Especially among the alliance of printers who saw themselves as artists rather than assistants to Linotype machines.

Most scholars identify the start of the Private Press Movement with a lecture given at the Arts and Crafts Exhibition Society in London on November 15, 1888. The timid speaker was Emery Walker, the . . . let's say, Yoda of that rebel alliance. Walker was a photoengraver, successful businessman, and vocal advocate of returning printing to its high-quality roots. He had been invited to present this lecture by fellow society member T. J. Cobden-Sanderson. T.J. can be Luke Skywalker, here, if you want—the printing Force was strong with that one.

Walker's address had a profound impact on the world of fine printing. What may seem like a grumpy old man's lecture about things "being better back in the day" was in fact a thrilling and inspiring call to arms. It sparked the creation of an entire community dedicated to examining every aspect of print, and finding a way to make the results as transcendent as possible. Most immediately, it inspired William Morris, the Pre-Raphaelite poet and textile designer, to start the Kelmscott Press, which specialized in limited edition, hand-pressed books of exquisite beauty.

Using a font he designed himself, and employing just two compositors and one pressman, Morris printed what has been called the first book of the Private Press Movement in 1891: his own novel *The Story of the Glittering Plain*. Morris's Kelmscott Chaucer, considered the masterpiece of that press, is often credited as a front-runner for the Most Beautiful Book in the World. Colin Franklin noted in 1969 that "[b]ooksellers take its current price for an index of the state of the nation," and that assertion is still pretty accurate in some circles today.

Meanwhile, Emery Walker presided behind the scenes, as he would for many of the most important printers of the Private Press Movement. Walker was the de facto silent partner behind Kelmscott; according to Morris's sometime private secretary Sydney Cockerell, "no important step was taken without [Walker's] advice and approval." When T. J. Cobden-Sanderson ventured into printing himself, his business partner was none other than Emery Walker.

The Private Press Movement lasted approximately from 1891 to the end of the Second World War. In all things, it was a rebellion against the impersonal mechanization of modern printing. It was a revolution of beauty. And if you were going to choose a revolution from the 1800s to throw your support behind, the one that results in artisanal books is probably better than ones that end in guillotines, political xenophobia, or ethnic cleansing.

The inspiration for this printing revolution was the books of Gutenberg's day, which were, in comparison to industrial age volumes, gorgeously fabricated and meticulously designed, from the viscous black ink and the carefully sculpted type to the hand-sewn bindings. Even in Gutenberg's day, though, machines were used to produce commodities on a mass scale. Herein lies the irony. Machines can't be all bad: the printing press itself is a machine. So how much mechanization is too much mechanization? The rebels who ran private presses rarely agreed among themselves on any particular detail.

One characteristic starkly separates fine press printing from mainstream printing: its ultimate goal. Whereas big publishers were out for profit, fine presses cared about making high-quality books—profits and nice clothing and warm houses and food be damned. The money made by presses such as the Doves Press barely factored into anyone's considerations because they cared about only the elegance of the final product. On this point both Walker and

Cobden-Sanderson were firmly in agreement. As a result, these presses created some of the most stunning books the world has ever seen, but it also made running a business difficult. Always short on cash, the Doves Press closed five years before Cobden-Sanderson's death. Its parts were sold off for just eighty-seven pounds—about 18 percent of its total start-up cost.

Even if Cobden-Sanderson could have foreseen the meager finances coming to him, he wouldn't have cared—at all. His wife, Annie, might have cared a little because she funded the Doves Press with her inheritance, but Cobden-Sanderson was a rebel with a cause. And that cause was gorgeous books. Or, as he described it, "The Book Beautiful."

Like other pieces of art that are sold as commodities, over the centuries books have been balanced precariously between "sacred vessels of Western culture" and "show me the money!" Both philosophically and monetarily, books are most often valued for what's written inside them. Still, we have a natural urge to make something physically beautiful in order to demonstrate its value to us. Some of us take that a bit further than others. For Cobden-Sanderson, money was entirely beside the point. Books themselves were divine: "This is the supreme Book Beautiful . . . a dream, a symbol of the infinitely beautiful in which all things of beauty rest, and into which all things of beauty do ultimately merge."

Good heavens, where can someone buy a piece of this sublime portable universe, marching ceaselessly through time and space? Answer: at the Doves Press sometime in the slim sixteen-year window of its operation. In October 1902 the Doves Press completed its original printing of John Milton's *Paradise Lost.* It was called by the famed English bibliographer A. W. Pollard "the finest edition . . . ever printed, or ever likely to be . . . I know no more perfect book in Roman type." He's really not exaggerating. You can expect to pay a few thousand dollars to obtain one today.

IN THE BEGINNING GOD CREATED THE HEAVEN AND THE EARTH. ¶ AND THE EARTH WAS WITHOUT FORM, AND VOID; AND DARKNESS WAS UPON THE FACE OF THE DEEP, & THE SPIRIT OF GOD MOVED UPON THE FACE OF THE WATERS. ¶ And God said, Let there be light: & there was light. And God saw the light, that it was good: & God divided the light from the darkness. And God called the light Day, and the darkness he called Night. And the evening and the morning were the first day. ¶ And God said, Let there be a firmament in the midst of the waters, & let it divide the waters from the waters. And God made the firmament, and divided the waters which were under the firmament from the waters which were above the firmament: & it was so. And God called the firmament Heaven. And the evening & the morning were the second day. ¶ And God said, Let the waters under the heaven be gathered together unto one place, and let the dry land appear: and it was so. And God called the dry land Earth; and the gathering together of the waters called he Seas: and God saw that it was good. And God said, Let the earth bring forth grass, the herb yielding seed, and the fruit tree yielding fruit after his kind, whose seed is in itself, upon the earth: & it was so. And the earth brought forth grass, & herb yielding seed after his kind, & the tree yielding fruit, whose seed was in itself, after his kind: and God saw that it was good. And the evening & the morning were the third day. ¶ And God said, Let there be lights in the firmament of the heaven to divide the day from the night; and let them be for signs, and for seasons, and for days, & years: and let them be for lights in the firmament of the heaven to give light upon the earth: & it was so. And God made two great lights; the greater light to rule the day, and the lesser light to rule the night: he made the stars also. And God set them in the firmament of the heaven to give light upon the earth, and to rule over the day and over the night, & to divide the light from the darkness: and God saw that it was good. And the evening and the morning were the fourth day. ¶ And God said, Let the waters bring forth abundantly the moving creature that hath life, and fowl that may fly above the earth in the open firmament of heaven. And God created great whales, & every living creature that moveth, which the waters brought forth abundantly, after their kind, & every winged fowl after his kind: & God saw that it was good. And God blessed them, saying, Be fruitful, & multiply, and fill the waters in the seas, and let fowl multiply in the earth. And the evening & the morning were the fifth day. ¶ And God said, Let the earth bring forth the living creature after his kind, cattle, and creeping thing, and beast of the earth after his kind: and it was so. And God made the beast of the earth after his kind, and cattle after their kind, and every thing that creepeth upon the

27

THE DOVES PRESS BIBLE, 1903–1905. *Image courtesy Heritage Auctions.*

Cobden-Sanderson next tackled the Bible. "In the beginning God created Life," he explained. "And it is this Life, this Life of each and of all of us, which in the language of the Press, we must COMPOSE." Between 1902–05, the Doves Press composed life.

The clear, stark beauty of this page feels like a marble statue translated from stone into paper. In person, it is breathtaking.

Anyone talking around you suddenly sounds like the adults in *Peanuts* cartoons, a dull, irritating *wah-wah-wah* next to the presence of the book in your hands.

This lofty rhetoric may seem a bit hyperbolic to people who view books as . . . just books. To the untrained eye, though, the simplicity is deceiving: "[The] masterly calligraphic initials, like the unforgettable opening of Genesis . . . were a perfect example of how to marry calligraphy and typography, and [the] setting was full of those almost invisible refinements that only another printer can recognize."

To Cobden-Sanderson and the revolutionaries of the Private Press Movement, books were the summit of artistic expression. What value had money when compared to the realization of humanity's true potential? It was dross, we tell you, shiny, metal dross. At least, that's how Cobden-Sanderson felt about it. His business partner had a more complex view.

Emery Walker may have been the Yoda whose speech sparked the Private Press Movement, but he was still a competent businessman. Two men can run a print business together, sure, but it's a bit harder if one is an artist with his head in the clouds and the other is a leader with a plan. Initially, it seemed that Cobden-Sanderson and Walker shared the same philosophy. For example, they both formally agreed to be paid in books. "[Cobden-Sanderson] and Emery Walker were simply to share the work, and, in return, each of them would receive a copy on vellum, and a dozen copies on paper, of the books they printed. If there were any profits after expenses, these would be divided equally." Profits were like getting to the end of a bag of Halloween candy. The valuable stuff has already been divided—vellum copies of *Paradise Lost* might be the full-size Snickers bars here—and all that's left are those awful rainbow-colored Tootsie Rolls and bite-size Milky Ways. Fine, take half of those. No one wanted them, anyway.

Both Walker and Cobden-Sanderson believed in producing beautiful books before profit. So far, so good. But pretty soon their differing philosophies rose to the surface. In particular, Walker believed he was an equal partner in the Doves Press. Cobden-Sanderson believed he was an idiot for thinking that.

Emery Walker may have brought some much-needed business savvy to the operation (along with a list of potential subscribers from the neighboring Kelmscott Press), but Annie had paid for everything, and Cobden-Sanderson was doing most of the work. The most serious dispute between the partners wasn't over profits. It was the fight on principle for ownership of the most important product to come out of their joint venture: the elegant, pristine, chills-inducing Doves Type.

BESIDES THE influence of technical designers and punch cutters, the Doves Type is a combination of two different typefaces dating back to fifteenth-century Italy, as well as Cobden-Sanderson's personal touch to the numerals. The Doves is a mirror of its unique time and place. While it reflects the reverence that printers such as Walker and Cobden-Sanderson paid to the pioneers of print in the fifteenth century, details have been subtly changed to make it as refined as possible to the early twentieth-century eye. It is the Private Press Movement in a nutshell.

What was true of the Doves Type is true of typefaces from any age: they are reflections of their culture. Even before print, the script we call Gothic developed in the Late Middle Ages as a way for scribes to write faster when the rise of universities increased the demand for books. In a time before paper came into common use, this script also produced writing in a more compressed form, to save on expensive parchment. In fact, it was originally called *textura* because the compressed style gave it a woven appearance.

Gutenberg's typeface in the first printed Bible was inspired by *textura* scripts.

We call typefaces like these "Gothic" because of the cultural shift that took place in the Renaissance. *Gothic* was "a term of derisive abuse which the Italian humanists applied to the traditional script of their contemporaries . . . [I]t was only meant to signify the classicist's contempt for the barbarism of those who neglect to follow the Roman models as the Goths and Vandals did." In other words, our modern name for this font family is a Renaissance-era insult aimed at the Middle Ages.

No matter the time or culture, our fonts change with us. In order to define themselves against the rest of the world, for example, early twentieth-century German patriots advocated the use of Gothic fonts. These fonts were deeply tied to major moments in the German literary tradition, such as Martin Luther's 1523 translation of the Bible. Maybe you can guess where this is going. With the Nazi Party's rise to power, Gothic fonts became part of the *Propagandamaschine.* Here's a real slogan from the time: "Feel German, think German, speak German, be German, even in your script."

In 1941, however, German script suddenly changed. Gothic type was labeled Jewish. "The type [was] being newly associated with the documents of Jewish bankers and the Jewish owners of printing presses." That may have been true, but there were other causes as well. According to Erik Spiekermann, there was actually a shortage of fonts in Gothic typefaces; moreover, conquered territories were having trouble reading it. In other words, Gothic became too impractical. What's the point in conquering a nation if the people can't read all the horribly racist posters you keep putting up? The dominance of Roman typefaces in German books today is partially thanks to the lack of printing supplies in the Nazi Party.

The typefaces of the Private Press Movement reflected the concerns and the culture of the period. William Morris's work at the

Kelmscott Press, which many people write off as simply *too much*, expresses a powerful printing philosophy. The pages are densely pressed, covered with elaborate woodcuts and a typeface almost too pretty to read comfortably. One fellow printer calls them "full of wine." This style was meant to be beautiful, yes, but it was also meant to create an "alienation effect," that is, an extreme statement meant to force a reaction. Morris wasn't simply suggesting his opinion; he was screaming it. It was this barbaric yawp that led the rebellion like a battle cry.

The moment of conception for the Doves Type probably occurred at a Sotheby's auction in December 1898. About a year and a half before, William Morris had died. By the following spring, his Kelmscott Press had closed, and the inventory was slotted to be sold off. In his journal from December 11, Cobden-Sanderson wrote about the need to create a typeface for "the Book Beautiful." Lucky for him, much of the movement's original inspiration was contained in Morris's collection of fifteenth-century books, which had just gone up for auction.

Two books in particular would have been of supreme interest to Cobden-Sanderson. The first was a copy of *Historia Naturalis* by the ancient Roman author Pliny the Elder, printed in Venice in 1476 by Nicolas Jenson. Jenson had created a Roman font for this book that has been described as "absolutely perfect," and was imitated for the next four hundred years. William Morris used Jenson's *Historia Naturalis* to create his first Private Press font, which he named the Golden Type. (Morris liberally altered Jenson's typeface, to make it thicker and more gothic. Cobden-Sanderson wanted a return to the thinner, cleaner look of the original.) From *Historia Naturalis*, the capital letters of the Doves typeface were formed.

The lowercase letters also came from a fifteenth-century Venetian printer, one Jacobus Rubeus, who printed *Historia del popolo fiorentino* in 1476. There is some debate as to how much access Cobden-

Sanderson had to the Jenson and Rubeus books while the Doves Type was being conceived, but this is where Emery Walker comes in. Walker had helped William Morris create his Golden typeface ten years earlier, by photographing and enlarging both *historiae*. Walker still owned those negatives.

One of Walker's photoengraving employees, Percy Tiffin, used this material to redraw a set of letters and numerals that would become the artistic model for the Doves Type. But Edward Prince, the punchcutter, had to carve them in 3D. This was no easy task. A reader might be wondering, how hard could it be to re-create those typefaces? The letters were there on the page in front of them. Just, you know, sculpt them again. A book's physical printing has an impact on how the typeface turns out, though. For example, how hard a letter is pressed into the page, or how much ink it has taken, changes the appearance of the type. Walker wrote that "nearly all of Jenson's [letters were] . . . over-inked and gave an imperfect view of the type." Prince had to "extract" their "true shapes" by hand from the various samples he was given.

Since the Latin alphabet doesn't have a *J*, *U*, *W*, or its lowercase equivalent, those letters also had to be added. (Fun fact: Latin did have a *w* sound, but it was attached to the letter *v*. So that famous phrase attributed to Julius Caesar, "Veni, vidi, vici," ("I came; I saw; I conquered") would have been pronounced, "Weni, widi, wiki"—which sounds like something a Roman toddler might say after subduing the Pontic Empire under his bed.)

Cobden-Sanderson took issue with the numerals that Tiffin created for the Doves Type, so he consulted an acquaintance at the British Museum and finally approved a set that looks like the room numbers hanging in every classy hotel built between 1910 and 1940.

The Doves Type can't adequately be called the creation of any one person. Two fifteenth-century printers provided the design of the letters; Walker owned the negatives and made revisions; Tiffin

1900 — 1915

NUMBERS PRINTED IN THE *DOVES PRESS CATALOGUE RAISONNÉ*, 1914.
From authors' personal collection.

redrew the letters; Cobden-Sanderson directed much of the operation and tinkered with the numerals; punch cutter Edward Prince transformed those model drawings into actual metal characters that could be sunk into matrices and cast into type.

One of the terms of Walker and Cobden-Sanderson's printing venture was that each partner should receive a font of Doves Type for his own use. By 1906, however, Cobden-Sanderson wasn't in much of a sharing mood. After finishing the print run on the Bible, he wanted Walker out, and offered him a sum of money to forfeit all claims to the Doves Type.

More than a few red flags had popped up before this, signaling that their partnership wasn't going so well:

- Cobden-Sanderson was nearly incapable of holding down a typical nineteenth-century job. Walker ran a thriving photoengraving firm.
- Cobden-Sanderson was a perfectionist with an artist's temperament. Walker was a level-headed businessman.
- Cobden-Sanderson was "done to death's door" by the amount of time he spent at the Doves Press. Walker checked in once a day, more interested in light management, without handling any actual presswork.

In a long letter written to Walker in 1902 (but never sent), Cobden-Sanderson outlined his grievances. When they started the press, they'd agreed to split the labor. Cobden-Sanderson would read the proofs, and Walker would oversee technical operations. It seems Walker wasn't living up to his end of the bargain. To make matters worse, he was critical of typographical errors and design choices that were supposed to fall under Cobden-Sanderson's jurisdiction.

Writing about their past projects, Cobden-Sanderson complained, "You objected to the spelling [in *Paradise Lost*], and you objected to the capitals in the text . . . to my arrangement of 'In the Beginning' and to the long initial I, and said 'It will never do,' you objected to the position of the title of the First Book of Genesis on the left hand page, and said it was 'hateful.'"

We're picking up a lot of angst here. By August 1902, Cobden-Sanderson wanted to dissolve the partnership completely. Since printing for the Doves Bible had just begun, he decided to bide his time. Three and a half years later, with the Bible behind them and his son, Dickie, working alongside him at the press, Cobden-Sanderson formally tried to buy out Walker's claims to the Doves Type. Walker was less than cooperative.

It would appear that, for sole ownership of the type, Cobden-Sanderson upped his original offer from one hundred to three hundred pounds. It would also appear that Walker was like, uh . . . no. Whatever amount Walker countered with was apparently too rich for Cobden-Sanderson, and the negotiations broke down for the next couple of months.

The following February, a mediation of sorts was reached between the two partners—or, at least, that was Cobden-Sanderson's impression. A mutual acquaintance drew up an arrangement that would allow for the end of the partnership in November 1908, and a payment of one hundred pounds to go toward the casting of a new font of Doves Type, if Walker so desired. He would later deny that he agreed to any of this. Which would have made any chitchat around the espresso machine particularly awkward over the next two and a half years, had Walker remained involved in the daily operations.

By this time, Cobden-Sanderson was in his late sixties. The sun rose and set on one thing in his world: fine printing. Everything else could take a long walk off a short pier, including personal acquaintances. "[Cobden-Sanderson] liked to watch plants and insect forms, to get new ideas for binding tools . . . he lost friends to whom his life looked peculiar. The new work became far more important than their small talk."

On one occasion, Cobden-Sanderson completely lost it when turning down the leather on a binding. In his own words, "I found it too short, and in a burst of rage I took the knife and cut the slips and tore the covers and boards off and tossed them to one side; then, in a very ecstasy of rage, seized one side again, tore the leather off the board, and cut it, and cut it, and slashed it with a knife. Then I was quite calm again; made so, I think by the wonder and awe and fear that came over me as in a kind of madness." Nothing odd about that—just re-creating the printer scene from *Office Space* because the headband on his binding was too short. It's not as if he were serially prone to rash and destructive behavior. . . .

To Cobden-Sanderson, handing over the most beautiful font that had ever been created so that Walker could use it for a second Doves operation outside his control was tantamount to Private Press blasphemy. *Here is the famed sword Excalibur. Should we use it to harvest wheat? Or perhaps chop firewood? Or maybe pick food out of our goddamned teeth? No sir, I tell you, no!*

Cobden-Sanderson gave his six months' notice to Walker in June 1908, which Walker promptly dismissed. Foreseeing the storm that was about to break, Cobden-Sanderson's son, Dickie, quickly abandoned the Doves Press to work for an uncle.

Emery Walker agreed that the partnership had to end. But while Cobden-Sanderson wanted to continue his fine printing crusade,

Walker thought the most prudent course of action would be shutting down the Doves Press, liquidating everything, and dividing whatever proceeds were left fifty-fifty. Considering what Cobden-Sanderson was willing to do to leather binding that was too short, we can only imagine the tour of human horrors playing out behind his eyes upon hearing that.

In December 1908, Cobden-Sanderson refused Walker further entry into the Doves Press. In June 1909, Walker initiated proceedings against Cobden-Sanderson in the High Court of Justice. On June 14, Cobden-Sanderson wrote, "[Walker's] proceedings at the utmost can only result in 'damages' or imprisonment: and to think of that! for nothing on earth will now induce me to part with the type . . . I have the will, and I have in my actual possession the punches and the matrices, without which it is impossible to have a Font of Type . . . I am, what he does not appear to realize, a Visionary and a Fanatic, and against a Visionary and a Fanatic he will beat himself in vain."

Two days later, their friend Sydney Cockerell brought them to the negotiating table. According to a new proposal, the partnership would be dissolved on July 23, and Cobden-Sanderson would be allowed to retain the Doves Type, to use at his discretion, for the rest of his life. Upon his death, however, the font would pass to Emery Walker.

So here you have two obstinate old men (Cobden-Sanderson, age sixty-nine; Walker, age fifty-nine) agreeing to a printing truce, each stubbornly attempting to outlive the other. It was basically the *Highlander* approach to font ownership. *There can be only one!*

Even as early as this truce, Cobden-Sanderson had secretly vowed that Emery Walker would never own his font, but its destruction wouldn't happen for almost a decade. Emery Walker left the Doves Press in the summer of 1909, and Cobden-Sanderson and his wife, Annie, continued running it for the next eight years. The press itself wasn't what you might call "profitable." It was now worth less than its original investment, and Annie's money, which was subsidizing the operation, had nearly evaporated. The printing house had to be closed and the press moved into the attic of the Doves Bindery. Annie and T.J. left their home so they could rent it for extra cash. In 1909, upon the advice of friends, Annie convinced her husband to begin printing the plays of Shakespeare. You know you're involved in a snobby movement when printing Shakespeare is considered selling out.

The Doves Press ran for a total of sixteen years, from 1900 to the first few weeks of 1917. It is worth noting that there was no perceptible drop in the quality of Doves printing after Walker's exit. For all the contacts and experience that he had brought to the operation, Cobden-Sanderson appeared to be the artistic visionary behind those Books Beautiful.

"It is my wish," Cobden-Sanderson wrote back in 1909, "that the Doves Press type shall never be subjected to the use of a machine other than the human hand . . . or to a press pulled otherwise than by the hand and arm of man or woman." In his "Last Will and Testament of the Doves Press," written in June of 1911, a solemn Cobden-Sanderson decreed, "To the Bed of the River Thames, the river on whose banks I have printed all my printed books, I bequeath The Doves Press Fount of Type."

Strangely enough, Cobden-Sanderson wasn't the only printer who sank his font into the waters of the Thames. In 1903, Charles Ricketts of the Vale Press threw his original Vale Type into the same river. More than forty years later, the Brook Type used by the Eragny Press (with whom Ricketts had formal involvement) was cast into the English Channel. Was there something in the water that drove fine printers mad, compelling them to sacrifice their font to it? As Ricketts explained, "It is undesirable that these fonts should drift into hands other than the designer's and become stale by unthinking use." This is a rather common fear of type designers. What happens when you're not around to guide its use? What if your beloved type ends up on a package of Charmin toilet paper? No one wants their baby to become the Toilet Paper Font.

Cobden-Sanderson's destruction of the Doves Type began just before Easter, on March 18, 1913. Over the course of three days, Cobden-Sanderson walked to the edge of nearby Hammersmith Bridge and hurled the punches and matrices of his type into the river. The loss of these components, used to cast pieces of Doves Type, would have made it nearly impossible to re-create any new sets of the font. After a short break, Cobden-Sanderson revived his plan of destruction in 1914, just in time for the madness that was about to spread across Europe in the form of the First World War. The brooding Cobden-Sanderson cited the war two years later as he was erasing the last remnants of his font: "If I am foolish, well, what can be more foolish than the whole world? My folly is of a light kind."

Knowing that the Doves Press would close forever at the end of the year, Cobden-Sanderson started the destruction of the font itself on August 31, 1916. "It occurred to me that it was a suitable night and time," he wrote, "so I went indoors, and taking first one page and then two, succeeded in destroying three. I will now go on till I have destroyed the whole of it."

The word *page* might be a bit confusing in this context. Cobden-Sanderson was talking about packets of metal type, each weighing over six pounds. Carrying twenty pounds of type the half mile between his house and the bridge in the dark, then heaving it over the side, would have been a moderately difficult feat for any seventy-six-year-old. When you envision a whole font of type, you might be picturing the total number of letters needed to fill up one of those old-timey partitioned printing cases. But when Cobden-Sanderson wrote that he was going to "go on till I have destroyed the whole of it," he was actually talking about 2,600 pounds of metal type.

In total, the Doves Type would have weighed more than a metric ton—and one very stubborn old man was determined to single-handedly destroy all of it.

At fifteen pounds (or two pages of type) per nighttime outing, Marianne Tidcombe estimates that Cobden-Sanderson took 170 individual trips to Hammersmith Bridge between August and January 1917 to do so. Sometimes he carried the metal type in linen bags, or wrapped in paper, but the most effective means was a converted wooden box with a sliding lid that he could overturn above the waters when he thought no one was watching.

"But what a weird business it is, beset with perils and panics! I have to see that no one is near or looking; then, over the parapet a box full, and then the audible and visible splash. One night I had nearly cast my type into a boat, another danger, which unexpectedly shot from under the bridge! And all night I feared to be asked by a policeman . . . what I had got in my 'box.'"

Ignoring the obvious potential for physical harm incurred from chucking fifteen pounds of metal type off a bridge into passing boats, Cobden-Sanderson seemed genuinely enlivened by the criminality of his act. This despite running the risk of a potential jail sentence. With a hint of glee, he remarked, "Hitherto I have escaped detection, but in the vista of coming nights I see innumerable possibilities lurking in dark corners, and it will be a miracle if I escape them all."

It was not so much a miracle as a question of no one caring about people randomly littering in the Thames. Before stepping forward to intervene, a Thames commissioner would have had to know all the details of Emery Walker's suit against Cobden-Sanderson in the High Court of Justice, as well as the resulting arbitration.

Over the five months that it took to destroy the Doves Type, no one had the slightest inkling what Cobden-Sanderson was up to—not even his wife, Annie. If he hadn't written with so much earnestness about his crime, the final whereabouts of the Doves font would have been lost to the ages. In addition to the sorts and the matrices and punches, Cobden-Sanderson burned most of the letters and other papers that had any connection to the Doves Press or the bindery. "I am determined that, as far as I am able to destroy, there shall be no debris left, no history of petty details, but only the books themselves."

The Doves Press officially closed its doors on January 21, 1917. T. J. Cobden-Sanderson survived his press by only a few years, dying in September 1922.

Exactly twelve days after his death, Annie was contacted by Emery Walker's lawyers and instructed to surrender the Doves Type, along with the punches and matrices, on pain of immediate litigation. In response, Annie published her husband's will, wherein he detailed his intention to destroy the Doves Type. A lengthy spat of legal proceedings ensued that likely cost Annie upward of twelve hundred pounds. But the destruction of the Doves Type had been effected, and no amount of money was going to bring it back. Walker even attempted to hire Edward Prince, the original punch cutter, to re-create the lost font. But by 1922, Prince was seventy-six years old, and neither his eyes nor his hands were up to the task.

Cobden-Sanderson's ashes were interred at the Doves Bindery in 1922. Annie's ashes followed four years later. In 1928 the River Thames flooded its banks and washed through the building, so it's likely that Cobden-Sanderson himself ended up at the bottom of the Thames, entombed in its waters right alongside his beloved type.

Emery Walker outlived his crusty old printing rival, but he spent the last ten years of his life knowing that Cobden-Sanderson had bested him. The Doves Type was gone forever. There was no hope of ever finding it again—unless, of course, you had some kind of Self-Contained Underwater Breathing Apparatus.

Cobden-Sanderson ultimately underestimated posterity's admiration of his font. Over the next century, one devotee of the Doves Type has proved to stand above all others: Robert Green, an English designer who spent three years painstakingly re-creating the type from books printed by the Doves Press. In November 2013, almost one hundred years after it was destroyed, Green released the first publicly available Doves typeface. Since this was the twenty-first century and no one needed punch cutters or metal matrices anymore, Green was able to release it as a digital download for a nominal fee.

In October 2014, while one of the authors of this book sat back daydreaming of finding the Doves Type in the Thames one day, Green went one step further. He hired divers from the Port of London Authority to scour the river for any sign of the lost metal type. Even after a century (and a few Hammersmith Bridge bombings by the IRA), Green's team succeeded in exhuming 150 individual pieces of the Dove font dumped by Cobden-Sanderson in 1916.

Now we have the Doves Type back (kind of). You can—nay, should—buy a digital re-creation of the complete font for forty pounds online. Doves Type even has its own Twitter feed. As a final act of justice, Green is donating half the recovered Doves font to the Emery Walker Trust. The execution of the original agreement between the Doves partners may have come a century late, but better posthumously late than posthumously never.

The Private Press Movement was a rebellion against the increasing industrialization of printing. It was a revolution fought with ink recipes, and carefully weighted homemade paper, and original typefaces inspired by the first decades of Gutenberg's press. From that movement emerged a "Visionary and a Fanatic" who would make such dazzling books that one glance is enough to induce goosebumps. Today Doves books are some of the most beloved and sought-after creations of the Private Press Movement. The spirit of that crusade led to a reexamination of every aspect of print. Its innovations spread throughout the printing business, becoming one of the strongest influences on twentieth-century book design.

Over the past century, typography and graphic design have blossomed in ways few could have predicted. But here's a prediction we're reasonably sure of: if the 1890s produced its Cobden-Sanderson in reaction to Koenig presses and Linotype machines, it's safe to assume that the twenty-first century will have numerous Cobden-Sandersons in reaction to the increasing digitization of our times. As long as humans stay human—and by that we mean rebellious visionaries—the spirit of the Private Press Movement will never sink into the dark waters of obscurity.

CHAPTER ELEVEN
11
BLIFTER!

ON A CLEAR DAY IN 1928, a plane flew over Manhattan leaving a trail of smoke reading, "MURDER?" It was the Roaring Twenties, the "Era of Wonderful Opportunity," and advertising was as drunk as the rest of us, Prohibition be damned. Exhibit A: The publisher Covici-Friede decides it's a good idea to emblazon a vaguely threatening but mostly confusing message across the sky as a promotion for the release of *Murder*, a book by Evelyn Johnson and Gretta Palmer. The Twenties blossomed in the form of movie tie-ins, gaudy dust jackets, barbershop quartet competitions, and "sandwich men" who functioned as walking billboards—all in the service of selling books.

Yet there was an imminent end to this literary carnival, where the excesses of exuberance flirted with disaster. Dragged into the Great Depression along with the rest of the world, by 1930 book publishers in America were facing an economic crisis—one that potentially spelled the end of the publishing industry as they knew it. Publishers needed someone to stand against the chaos and hold everything together. They needed Eddie Bernays, known to the advertising world as the Father of Spin.

IN ADVANCE of the carefree 1920s, the printing industry had taken giant leaps in mechanization. Innovations such as the Linotype machine and high-speed rotary presses enabled publishers to churn out books and periodicals at many times the rate of hand presses. The Industrial Age brought the world a flood of books. Yay! But that wasn't actually a good thing. No! At least not yet. Um . . . ambivalence? As print historian Ted Striphas has observed, "Today accumulating printed books and shelving them in one's home may seem like mundane facts of life. In the first decades of the twentieth century, however, those activities couldn't be assumed and needed to be learned."

Learned? Why would someone need to learn to buy books? That's just what you do whenever you go into town, or get birthday money, or fill the swear jar all the way. But this consumerist mind-set that we take for granted owes a great deal to folks like Eddie Bernays. At the turn of the twentieth century, most Americans simply didn't buy books. One in three homes owned no books whatsoever. Those that did usually limited them to a Bible, an almanac, and perhaps a few spellers for their children. What other books did a person really need? In 1921 only 4 percent of Americans visited a bookstore, and almost all of them lived in metropolitan areas.

In the nineteenth century, most books were either a luxury beyond the average American's financial reach or "like furniture . . . hand-me-downs or the rare gift for a special occasion . . . there was little need to buy another [book] until the old [one] wore out." No wonder people didn't buy books. If we had to read *The Da Vinci Code* until it literally fell apart before we could buy the next *Song of Ice and Fire* we'd just stop reading altogether. Rabid nineteenth-century readers were more likely to simply subscribe to their local circulating libraries.

The printing industry in America had reached an unprecedented level of efficiency, but our buying habits had not. Manufacturers such as Henry Ford had learned how to run their businesses as profitably as possible with the help of the assembly line, and the book world followed suit. Publishing, however, had become almost too efficient. According to Striphas, "The book industry had, in a sense, become a victim of its own success. Its capacity to produce books had grown so rapidly and to such a degree in the early twentieth century that it had lost touch with supply and demand—if it ever had it to begin with."

That we haven't always been the responsible Black Friday consumers we are today may be a concept difficult for most Americans. It's only recently that we've been convinced we have to buy things on a regular basis in order to be contributing members of a growing economy and people with individual worth. How can those around us possibly understand that we have a personality or basic taste unless we own things to prove it? Americans needed to undergo a fundamental reeducation. We needed to learn that what we buy is every bit as important as what we say or what we do.

> Consumptionism is the name given to the new doctrine; and it is admitted today to be the greatest idea that America has given the world; the idea that workmen and masses be looked upon not simply as workers and producers, but as *consumers.*
>
> —Christine Frederick, *Selling Mrs. Consumer*, 1929

The greatest thing our country has given the world is consumption? That . . . that doesn't sound right. Yet, in the twentieth and twenty-first centuries, consumerism in America became one of our defining traits.

As the trend toward consumerism grew in the early years of the twentieth century, publishers were poised to take advantage of a number of cultural changes that spurred interest in private reading. For one, we stopped reading together. Picture a cozy Victorian scene, where a family gathers around the fireplace and reads aloud from the Good Book. While Dad or Mom dutifully droned on, the kids would be trying to find anything (please, God, anything at all) to focus on that was more interesting than scripture. But about the turn of the twentieth century, solitary reading gained more and more popularity. Most appealing were the kinds of books that you don't really want to read together with your dear old dad, aka novels.

The landscape of American life was also changing, and literary life changed along with it. Illiteracy had declined dramatically in the early decades of the twentieth century, and urbanization had ballooned. As any New Yorker will tell you, cities are particularly great for providing bookish resources: libraries, bookstores, schools, and universities. So as Americans moved into the cities, they read more books.

When Americans went to war, they also read more books. Down in the trenches of the First World War, books served as cheap, quiet entertainment amid the hurry-up-and-wait practicalities of the average soldier's life. Publishers of the early twentieth century agreed that the most successful American ad campaign in recent memory had been the government's push for donations of books to send to World War I soldiers. "[The] sign 'Books Wanted' appeared in the newspapers, on millions of posters, on cards, on streamers and banners, in street cars, on billboards, on the backs of books—wherever there were eyes to behold." The success of this unprecedented national campaign led publisher Alexander Grosset to boast in 1919, "At last the book business has come into its own."

Now all publishers had to do was take advantage of all those potential readers—which wasn't as easy as it might seem. "I hold very strongly to the opinion," said one New Jersey librarian, "that book publishers are a very stupid lot as far as the art of salesmanship is concerned." In fact, advertisers were doing ten times better with Abba-Zaba bars and Bit-O-Honey than they were *Tales of the Jazz Age*. "The American Academy of Political and Social Sciences . . . estimated that the average American [in the 1920s] spent $1.10 for books annually, compared with . . . $11.00 for candy."

Helen Woodward, one of the best-known advertisers in the trade, argued in 1920 that it was time "publishers invest in an author with the same foresight which a soap manufacturer might invest in soap." It was in this decade that advertising agencies started offering services tailored specifically to the book industry. Print ads devoted to the author himself, rather than the work alone, began to appear. Author portraits became more common on dust jackets. Blurbs, a term invented only in 1907, began their hyperbolic ascent. Where there used to be blank pages at the end of a book, now Doubleday, as a teaser, printed the first chapter of the next book in the series.

Today many people associate the earliest dust jackets with famous books from the 1920s. Though this is far from accurate (you can find dust jackets on some mid-nineteenth-century books), the Jazz Age is the time when jackets came into their own as reliable tools for promotion. Just think of that iconic jacket for *The Great Gatsby*—the saturated blue, the piercing green eyes, the yellow spark above the author's name. This jacket, issued in 1925, now spells the difference between a $4,000 first edition and a $150,000 first edition. Dust jackets were that blast of color appearing in every store window—flashy ads for the books that publishers wanted consumers to buy. They were so successful, in fact, that some critics worried jackets might be mistaken for real art: "[W]e predict that you will live to see Bibles with a jacket, by Billy Sunday." (This prediction did come true—you can find Bibles with dust jackets today—though as far as we can tell, that had little to do with the efforts of evangelical preacher Billy Sunday.)

Not everyone fell for the hype. In 1919, Yale professor William Lyon Phelps grumbled: "I wish the publishers would quit putting jackets on their books. The first thing I do when I get a book is to throw the jacket away without reading it. It has no business there . . . To find laudation and praise on a slip-cover often antagonizes intelligent readers. The slip-cover should have the name of the book and author, and a sober statement of the purpose of the book."

Happily for everyone, Phelps wasn't taken seriously. And now, because original dust jackets are so valuable to collectors, Phelps's stripped-down twentieth-century collection would be worth next to nothing. With the exception of books from the earliest years of the century, most of these first editions aren't considered collectible at all if they are missing their dust jackets.

Publishers were beginning to figure out the concept of publicity. One infamous printer of paperbacks, E. Haldeman-Julius, realized that his reprint of de Maupassant's *Tallow Ball* wasn't selling so well. *Tallow* and *ball* are not particularly interesting words. But do you know what is? *Prostitute.* Haldeman-Julius reissued the book under the title *A Prostitute's Sacrifice*, and sales went from fifteen thousand to fifty-five thousand copies.

The aforementioned jacket-stripping Phelps entered the fray when he released a list of the "100 best novels." Because it's a list, and it's opinionated, and it's about books, you know it's going to be hella controversial. Soon everyone was talking about it. Thus Phelps participated in what might be called the inaugurating event of BuzzFeed (a website he surely would have despised). Literary discussions were becoming publicity stunts, and bestseller lists were becoming news.

One of the most natural advertising tools for a book was the perception of status, prestige, and power bestowed upon its owner. To do that most efficiently, we needed a way to physically show people around us that we were better than they because we'd read *Ulysses*. (Or at least had purchased *Ulysses* and read the first thirty pages, then skipped to Molly's soliloquy at the end.) Enter the bookshelf.

"Introduce a little touch of modernism," announces one article from the March 1929 edition of *American Home*. "A moderately worn appearance lends flavor to a book . . . If you want your books around you, you must have proper receptacles for them. While the covers of the books may be ever so worn, if they are attractively housed, the effect will be pleasing."

The bookshelf! Where American consumerism and neurotically obsessing over your social status can be displayed in one convenient location. The article goes on to declare in Stepford Wife–speak, "Certainly you, yourself, will be far better satisfied when surrounded by your old favorites than if you had a most harmonious array chosen solely . . . to please the eye but quite devoid of anything within." A bunch of glass grapes on a doily may be pretty, but like you, they'd be culturally dead inside.

Publishers loved this development. Let's face it, nothing is sadder than an empty bookshelf. According to Janice Radway, "Potentially every book sale could generate two forms of profit. On one hand it could generate cash for its publisher. On the other hand it could also produce perceived changes in the status of the individual who bought it." This is known as the "fetishism of commodities," that is, investing objects (in this case, books) with "certain naturally occurring, inherent properties." For example, consider the self-confidence one might achieve from owning a set of Emerson essays, as this ad recommended: "To fully recognize what magnetism there is in your own personality read CULTURE, WEALTH, BEHAVIOR, POWER."

With Americans building and filling their bookshelves to keep up with the Joneses, the future of the printing industry should have looked bright—and indeed it did, at least until the fateful year 1929.

You know what American economic disaster we're talking about here, so say it with us: the Great Book of the Month Club.

The Book of the Month Club actually launched three years earlier, as the brainchild of New York adman Harry Scherman, but 1929 was the turning point. It had taken him years to perfect a subscription-based service that sold books using methods similar to those employed by companies selling canned goods, tobacco, and soap. It's also how meth is sold today (sort of): "[Scherman] realized that he could not make money unless, following their first purchase, buyers were hooked into returning for additional ones. After some false starts, he devised a plan that combined the use of the mails with a subscription feature that insured the necessary 'repeat business.'"

By 1929, with a monthly cycle of guaranteed sales numbering in the tens of thousands, the Book of the Month Club had accumulated real weight to throw around with publishers. In the wake of its success, other book clubs soon sprang up, and many of them didn't play well with others. In particular, a subscription service known as the Literary Guild attempted to sell copies of new hardcover fiction at dramatically slashed prices. The Literary Guild was a bit like the Incredible Hulk to the Book of the Month's Bruce Banner: a good idea gone way, way out of control. While the Book of the Month Club relied on the taste and celebrity of its well-respected literary judges, in 1929 the Literary Guild began emphasizing cheap prices as its main selling point. Book club editions flooded the market, and even today they are the bane of collectors, who have all had the experience of finding an apparent first edition of *The Grapes of Wrath*—in dust jacket!—only to learn that it's a book club edition worth ten dollars, not ten thousand.

Publishers had experimented with price cutting in the past, but it was a bit like poking a sleeping bear. Despite browbeatings from the *New York Times* and other venues that wanted to see a universal *drop* in prices, new books were rarely considered a practical sell at only a dollar a piece. Publishers argued, on the contrary, that price *increases* were justified. According to their numbers, the cost of making a book—the cloth, paper, wages, etc.—had increased 75–80 percent over the past few decades, while book prices had gone up only 50–60 percent. Now the Literary Guild comes along and undercuts the whole trade. To make matters worse, major department stores started selling price-slashed books as loss leaders in order to increase customer traffic. In 1929 the American publishing industry was threatening to buckle in on itself as thoroughly as the national economy around it.

The success of monthly book clubs was one of the major factors leading to a small miracle: publishers coming together to cooperate. Those who choose to make their living by creating and selling books naturally tend to be a bit rebellious. This much cooperation among publishers during the price wars of the 1930s was almost unheard of.

Of course, there wasn't full agreement across the trade. Eddie Bernays wasn't called upon by a unified front of publishers to undermine book clubs. He was called upon by a group of established publishers to undermine a different group of publishers: young upstarts who had begun price cutting on their own in order to compete with the clubs. Simon and Schuster, Coward-McCann, and Farrar and Rinehart (all three energetic and ambitious publishing houses founded in the 1920s) announced that they were reducing new fiction hardcovers to only one dollar. Doubleday, Doran, made up of two older publishers that had merged in 1927, broke ranks and followed suit. This posed a potential disaster for the book industry.

You could find books for a dollar (or less) before May 1930, but the new price-cutting strategy wasn't focused on reprints, or remainders left over in warehouses. They weren't cheap paperback Westerns or mysteries that you could buy at your local cigar or drugstore, either. These were first-run hardcovers—for a dollar. That would be like AMC, Regal, and Cinemark getting together to fight Netflix by offering day-one new releases at 75 percent off.

In an industry facing slumping sales since 1928, a subscription service hijacker (or three) since 1926, and an extreme new counterstrategy that might sink American publishing, someone needed to act. Someone needed to fundamentally rearrange the public's buying habits. People had to be convinced to read more, and then further convinced that paying full price for books was the most moral, the most patriotic, the most copacetic thing a citizen could do. Get Eddie Bernays on the horn.

EDWARD BERNAYS was the son of Jewish-Austrian immigrants who brought him to New York when he was just a year old. He had two sisters, both of whom stayed behind in the Czech Republic for a time in the care of his internationally famous uncle Sigmund Freud. Edward's father envisioned his son carving out a career for himself in the field of agriculture, "because he believed that America's future rested on the development of its rural areas." But even as a young man, Edward was more like his uncle. He wanted to get into people's heads and change their behavior. He wanted to "pull the wires which control the public mind."

With those life goals, the Father of Spin could easily have been a psychoanalyst himself, or a politician, or a supervillain. (A supervillain called the Father of Sin. *You don't want to pay full price for* The Story of Doctor Dolittle*? KAPOW! Now you do! Mwa-ha-ha-ha . . .*) But none of those professions would have been as financially lucrative as convincing Americans to buy things. At the height of the Great Depression, Bernays was raking in more than $98,000 a year, which equates to over $1.5 million today.

Before 1914, Eddie referred to himself as a press agent. Just after opening his new firm in 1919, he began using the title "public relations counselor." What was a public relations counselor? Well, it was part psychoanalyst . . . and part politician . . . and part supervillain. In 1928, he was brought onto the "Reach for a Lucky Instead of a Sweet" campaign, which tried to convince women to smoke cigarettes instead of eating dessert. As it turned out, preying on women's physical insecurities worked. (Someone should really let today's advertisers know about this little trick.) Bernays later admitted, "I didn't like the taste [of cigarettes]. I prefer chocolate." Convincing people that, against all reason or logic, they should prefer the taste of tobacco to that of chocolate is pretty much the definition of a supervillain.

"It is his capacity for crystallizing the obscure tendencies of the public mind before they have reached definite expression which makes [the public relations counselor] so valuable," Bernays asserted. More valuable to a company and their products than lawyers, even. "[I] decided that public relations advice is more important than legal advice, because legal advice is based on precedent, but public relations advice might actually establish precedent."

In 1923, Bernays authored *Crystallizing Public Opinion*, the foundational work on public relations that is still being used in classrooms and boardrooms today. It showcased the power of a corporation's or government's ability to employ mass psychology to manufacture public opinion. Could this also be termed propaganda? Propaganda, schmopaganda. "The only difference between 'propaganda' and 'education,'" he wrote, "is in the point of view." Call it whatever you want, Bernays was good at his job. With his three-piece suit and impressive track record, he looked and acted the part of the slick New York PR man.

When American Tobacco asked Bernays to counter the taboo of women smoking outside (because, "damn it," he was told, "if they spend half the time outdoors and we can get 'em to smoke outdoors, we'll damn near double our female market"), Bernays suggested consulting American psychoanalyst Dr. A. A. Brill. "He might give me the psychological basis for a woman's desire to smoke, and maybe this will help me."

"Some women regard cigarettes as symbols of freedom," Dr. Brill told Bernays.

Okay, sure, breaking down taboos and all that.

"Smoking is a sublimation of oral eroticism; holding a cigarette in the mouth excites the oral zone."

Right. Psychoanalysts and their phalluses.

"Further, the first woman who smoked probably had an excess of masculine components and adopted the habit as a masculine act."

In other words, liking freedom (and cancer and fellatio) in 1928 was apparently a "masculine component." Bernays's solution was to "break the taboo against women smoking in public" by turning it into a feminist act. He even set up a march touting cigarettes as "torches of freedom."

And it worked. The parade garnered national attention, and the taboo of women smoking in public did begin to shift. "Age-old customs," Bernays wrote, "could be broken down by a dramatic appeal, disseminated by the network of media."

This was the most recent body of Bernays's work that publishers looked to when, in the summer of 1930, they went searching for someone to save their industry. After all, if Bernays could change the public perception of smoking, if he could convince people that tobacco was better than chocolate, then maybe he could work his same magic on books.

THE COALITION that hired Eddie Bernays was headed by such giants as Alfred A. Knopf and Henry Hoyns of Harper and Brothers. They pooled their resources and enlisted

Bernays for a three-month PR blitz. After doing some preliminary research, Eddie formulated a two-pronged strategy: "The first was to convince the public and the price-cutting publishers that dollar books were not in the public interest. The second was aimed at increasing the market for good books . . . [O]nly in this way could book publishing become a stabilized, profitable business."

Amid these challenges thrived a parasite that many publishers agreed was pure "evil," rotting the industry from the inside out. This consummate embodiment of villainy was none other than "the book borrower, the wretch who raised hell with book sales and deprived authors of earned royalties."

At least with dollar books, folks were making a dollar. But if you were the kind of no-good flimflammer who *borrowed* books, you were considered a swindler consciously burgling American printers and publishers and authors and booksellers out of their rightful dues. Bernays thought these literary clouts were enough of a plague that he initiated a campaign to shame them into buying books like decent Americans.

Because "events usually gain more attention than a statement," Bernays and his firm "launched a nation-wide search for a lethal epithet" that would make people think twice about ever borrowing or lending books. This campaign for a bookish scarlet letter took the form of a national contest, with entries pouring in from all over the country. Some of the finalists were *Book Weevil*, *Culture Vulture*, *Bookbum*, *Libracide*, and goddamn it, *Bookaneer* (truly an inevitability wherever books and theft are concerned).

The winner of the contest was Paul Stoddard, a high school English teacher from Hartford, Connecticut, who came up with the shame moniker *Book Sneak*. As a reward, he was presented with fifty books. *Oh, good*, Mr. Stoddard must have thought. *Who needs a cash prize during the Great Depression, anyway?*

Now that he had found an appropriate bibliocurse, next on

Bernays's list was convincing the public that dollar books would bring about the total ruination of American publishing. To accomplish this, he formed the Book Publishers Research Institute. This sounds all fancy and official, which was exactly what Bernays was counting on. According to the quasi-scientific findings of the BPRI, "The profits of book publishing were so small that the dollar book would cause the economic death of six thousand book retailers." If we, as a nation, continued down the poisonous road of cheap books, publishers would be forced to resort to unsavory discounting, thereby dropping the quality of printing, demoralizing authors, and bringing the fifty-million-dollar book industry to its knees. *Blifter!* That was another entry, which is presumably the combination of *book* and *lifter. Blifter!*

In addition to the BPRI, Bernays publicized studies done by the U.S. Census Bureau and the Department of Commerce showing that 76 percent of all books being read were fiction. Women read books solidly year 'round, while men spiked during the winter. According to internal findings, Americans in 1930 read, on average, one book per year. This was a "sad figure," according to Bernays, and one that desperately needed to change.

Bernays launched initiatives to convince the public that reading was a worthwhile activity, especially if you were male. The BPRI conducted studies of highly successful "industrialists" and claimed that, on average, these influential men read two hours per day (even in the summertime *gasp*), devoting themselves mostly to nonfiction topics such as politics, history, and economics. Journalists at the time questioned the BPRI's findings, but it didn't matter. The statistics being published implied that you could also become an accomplished, affluent industrialist if you just read more. (For two authors who have managed to make a living largely based on how much we read, we do have a hard time disagreeing with this.)

And hey, habitual reading could even make you president of the

United States (or at least more like him). "We did not limit our research of reading habits to the living; we also explored the past . . . Benjamin Franklin was mightily influenced by [Cotton] Mather's essay 'To Do Good.' Thomas Jefferson favored a wide range of reading; John Quincy Adams spent much time studying books . . . [and] John Adams advised his daughter Abigail to spend her solitary hours in reading so that she might better be able to attend to the education of her children. Even the career woman, a relatively new phenomenon at that time, was studied . . . [B]ooks played a potent role in her life."

As early as 1856, Walt Whitman had pulled a similar stunt, stamping onto the spine of the second edition of *Leaves of Grass* an excerpt from his personal correspondence with the much more famous literary figure Ralph Waldo Emerson: "I greet you at the beginning of a great career." This was essentially a blurb. Some have even called it the first instance of modern American book advertising.

While nonfiction titles made a respectable showing during the 1920s, in line with Bernays's marketing push, they truly outdistanced fiction in popularity in the 1930s, ending the long-established dominance of novels. It's no surprise that the Great Depression sent people running toward books about business, leadership, religion, self-help, and even frivolous entertainment such as bridge or crossword puzzles. The heyday of F. Scott Fitzgerald was over. (Someone probably should have mentioned this to Fitzgerald, since he considered his 1934 novel, *Tender Is the Night*, his masterpiece. It also has a killer dust jacket design, by the way.)

According to Bernays's BPRI, if you want to be successful, read a book. If you want to be more successful, read more books. If you want to be successful and also not a bibliofelon fink bastard, pay full price for the books you read. To hammer this point home, Bernays laid bare to the public the profit margins of the publishing industry. From the two dollars that came from selling first-run novels, thirty cents went to the author. Printing ate up another thirty-seven cents. Overhead, advertising, and promotion accounted for a further forty-five cents. Add in retailer discounts, and the grand total profit on a newly released book was around a nickel. Unless three thousand copies of the book sold, there were virtually no profits at all.

Then Bernays publicly crunched the numbers for dollar books, showing that it was impossible for the industry to survive much longer. "We kept hammering away at this book-keeping on books, and our efforts evoked favorable comment in many media. It was obvious that the book business could not run successfully on this price structure."

Greader! That was another entry. *Greedy reader*, maybe? Or *grifting reader*?

From the summer of 1930 to the following December, Eddie Bernays mustered all his psychoanalytic tricks to help change the way Americans perceived books. Publishers marketed books as physical talismans capable of bestowing wealth, prestige, and power upon their owners. Bookshelves swung into fashion in order to display those talismans and leave us wanting more. America marched gleefully down the road of consumerism. In this grand new utopia, where anyone could become president (if you were white and male) just by reading books, the worst thing you could be was a *Viperous Volume Vulture*, or a *Greeper*, or a *Borrocle* (good luck unpacking that one). If we all worked together, we could save our publishing industry. And maybe, just maybe, save ourselves in the process. By paying full retail price for *The Maltese Falcon*. (A book whose poor sales, incidentally, were blamed on readers not knowing how to pronounce the word "falcon," and thus afraid to ask for it at the bookstore.)

"Our campaign registered full success," Bernays boasted in his memoir, "On December 19 . . . the New York *World* carried a front-page double column headline: DOLLAR BOOK IDEA IS ABANDONED . . . The article said, 'That bright child of literature—the $1 book—has been left shivering on the doorstep.'"

Well, not exactly. Like many admen, Bernays was most skilled at selling himself. The price war was perhaps the single biggest controversy of the following decade. One side would claim a victory (like Bernays) and within months find itself retreating. It was the publishing world's Stalingrad, only one fought with ads instead of bullets and mortars. In 1934 the American book trade accepted a code from the National Recovery Administration (NRA) that limited the types of price cutting a publisher could do (whew, sigh of relief)—until 1935, when the NRA codes were declared unconstitutional. Damn you, the NRA!

As with the Depression, the biggest turning point for the publishing price wars was not clever writing or flashy campaigns; it was America's entry into the Second World War. And as you might recall, war is good for books. Nope, nope, we'll rephrase that. War is good for increasing reader statistics within soldierly demographics.

In the end, Bernays's focus on the health of the book industry contributed to another major turning point in American publishing: the Cheney Report. Working together yet again, publishers hired O. H. Cheney as the director behind a landmark economic survey of the industry meant to diagnose the problems obstructing their profits. In this case, even more publishers and booksellers cooperated to pay for the survey.

Cheney was an outsider. A very qualified outsider, but a publishing outsider nonetheless. When his 150,000-word report was issued in 1932, it didn't spare any feelings. (One historian describes it succinctly as a "bomb.") The practices of publishers were raked over by a neutral third party, and everyone came away with egg on their faces. "Cheney blasted publishers and booksellers for relying on intuition . . . rather than operating on a scientifically sound, statistically driven 'fact basis.' He disparaged publishers and editors for their lack of creativity in developing the talents of first-time authors and scolded them for 'murdering' potentially successful titles by releasing them into a field already so overcrowded that they simply 'cannibalized' one another." Many were so offended by the tone of the report—*murder? cannibalism?*—that they refused to discuss it further. Thankfully, some wiped the egg from their faces and got to work—decades later. Emphasized by Cheney's Report in the 1930s, the idea of using industry-wide standardization to increase efficiency finally achieved widespread acceptance around the late 1960s, when the revolutionary ISBN system was introduced. Now all merchandise could be coded using a universal standard that would communicate information about a book across publishers, bookstores, and libraries, making it easier than ever to regulate, buy, and consume books.

As bibliophiles in the 1920s and '30s already knew, books were more than the paper, ink, cloth, and leather that made up their parts. They were the physical expressions of our true potential. If we wanted to be romantic or witty, we needed to buy F. Scott Fitzgerald's *This Side of Paradise*. If we wanted to feel masculine angst, we needed to buy Ernest Hemingway's *The Sun Also Rises*. If we wanted to see glamour and scandal in the world around us, we bought Edith Wharton's *The Age of Innocence*.

In the 1920s, booksellers realized they were selling culture just as much as physical books. "[Individual b]ooks cannot be advertised, but reading can be," announced one adman. Looking at our own bookshelves today, we're going to say they succeeded.

With the twenties roaring recklessly toward the Great Depression, consumerism must have seemed an inevitability. This waking giant would define us as Americans and, in 1929, undo us as Americans (also 1973, 1982, and 2008). Surely, publishers thought, there had to be a way to make this behemoth serve the interests of the printing industry. Entrepreneurs such as adman Harry Scherman knew it could be done. Start a club, a monthly club, where folks can have their books selected and then mailed right to their door. Then create a customer-retention strategy that combined a pyramid scheme with strategies for selling heroin, and—voilà!—a whole country chasing the literary dragon.

In the case of Eddie Bernays, sometimes the crystallizing power of advertising could help save America's printing industry. Other times, it convinced people that getting lung cancer was liberating and tasted better than chocolate. Sometimes it made us believe that life would be better if we bought and read Ralph Waldo Emerson. Other times, it convinced us that something such as the McRib, the Hot Dog Bites pizza, or a jar of Baconnaise was an actual food that human beings should put in their mouths. But that's the price of American consumerism. You can't have *As I Lay Dying* without Baconnaise. (In fact, one pretty accurately describes the other.)

Even the Father of Spin realized the potential for advertising running amok. This was demonstrated by one memorable event at a dinner party in 1933. Bernays and other dinner guests were seated under a large elm tree discussing the issues of the day, which in 1933 were Hitler and the rise of the Nazi Party. This was a time when opinions were split on the rising German leader. Some guests "wrote him off as a crank who wouldn't amount to much." Karl von Wiegand, a foreign correspondent who had just returned from interviewing Joseph Goebbels, was also in attendance. Goebbels had shown Wiegand his "propaganda library," as Bernays wrote, which was the "best Wiegand had ever seen." Sitting there in the library of the master propagandist of the Nazi Party had been none other than *Crystallizing Public Opinion*, by Austrian American Jew Edward L. Bernays.

"This shocked me," Bernays wrote, "but . . . obviously the attack on the Jews of Germany was no emotional outburst of the Nazis. [It was] a deliberate, planned campaign."

I guess we're lucky that American PR men in the 1930s were more focused on selling full-priced copies of Steinbeck's *Cup of Gold* and less focused on selling international crimes against humanity. In hindsight, it made that sticker price of $2.50 just a little easier to swallow.

CONCLUSION
NOTHING MORE DECEPTIVE THAN AN OBVIOUS FACT

"PEOPLE NEED BOOKS," CHRISTOPHER MORLEY says, "But they don't know that they need them. Generally, they are not aware that the books they need are in existence." With so little space and so much to say, we've written this book as the academic equivalent of a tailgate party. We are the booze-soaked, hot wing–stained bash before the big event—which in this case would be sipping a cup of tea while quietly reading a biography of Thomas More.

In the world of books there are endless roads to explore. There are curiosities and triumphs, comforts and provocations. The twists and turns are as endless as the peculiarities of the human mind, for we have translated our inner universes onto the printed page. The single artifact that best captures the human spirit is the book—which means that it inevitably captures both our glory and our frailties.

While writing this book, we came across a 1593 painting of Thomas More, our Tree of Truth villain from the English Bible chapter, sitting with his family. This painting is a copy of a lost work by Hans Holbein, one of the great portraitists of the Renaissance. The original is missing, but Holbein's prepainting sketch did survive. It shows More's wife and two other women kneeling on the floor in front of the famed British statesman. This did not sit right with More. As in, his wife was literally not sitting right. Rather than kneeling on the floor like a servant girl, she should be seated on a chair like a normal human being. Even More's jester is sitting on a chair, for heaven's sake. Shouldn't the lady of the house have at least as many rights as a professional fool? The 1593 copy of the finished Holbein oil portrait shows that, in the end, More got his way. His wife is indeed seated next to him. Well, seated a foot lower than him, but still, very PC for the day.

"There's nothing more deceptive than an obvious fact," Sherlock Holmes once observed. Thomas More was more than a snarling, frothing attack dog obsessively tearing at Protestants during the early upheavals of what became known as the Reformation. He did all that, no doubt, but Rottweilers can have cuddly feminist sides, too. Rottweilers can be loyal. Rottweilers can be protective. Rottweilers are more complex than one might believe if all one knew about them was their proclivity for removing the skin from trespassers' skulls. Or burning people alive.

In less metaphorical terms, Thomas More was human, with complex motivations and inner contradictions. We'd say he was just like us, but we expect a bit more from our statesmen (and Catholic saints) today. So let's say he was like other people from his time (who were white, and male, and wielded a staggering degree of political power). Actually, he was one of the better people of his time. Did he say someone should shit and piss into Martin Luther's mouth? Yes. Technically he wrote and published this all over Europe. Was More also forward thinking when it came to women? Did he love his children? Was he well regarded by friends? Was he a master politician? Was he pious within his own code? Was he an exemplary husband? A brilliant author? A formidable philosopher? The greatest mind of English humanism? A tragic martyr? Yes, to every one of those. He was even considered an exceptionally funny guy for the Renaissance; one scholar calls humor "integral to More's purpose" in his *Dialogue Concerning Heresies* (the same work in which, we can't help but note, he uses a parable of Jesus to justify burning heretics).

Could a historian write a chapter on Thomas More that made him the hero of the story? More was canonized in 1935 and declared the Patron of Statesmen and Politicians by the Vatican in 2000, so yes. Absolutely.

When choosing the subtitle for this book, we elected to use *Irreverent Stories from Book History*, rather than *The History of the Book*. After writing a work that attempts to highlight a few events (so very few) from half a millennium of history, and knowing firsthand the ocean of material that could never be included, we find that this distinction has proved crucial. While we hold profound respect for Edward Gibbon's *The History of the Decline and Fall of the Roman Empire*, it's technically not accurate. Gibbon's prolific volumes are *a* history of the Roman Empire's fall, as told by Edward Gibbon. Thomas More's depiction in the "Trees of Truth" chapter is *a* depiction: as factually accurate as we can make it, but certainly not exhaustive with regard to the nuances of More's conflicted and endlessly fascinating personality.

The writer and philosopher George Santayana coined the historian's favorite aphorism, "Those who cannot remember the past are condemned to repeat it." It may seem at times that people living three, four, five centuries removed from us have little to contribute to the issues of our day. Yet, in many ways, the problems we face now are iterations of the past. Especially as the human race blazes its way through the technological advancements of the digital age.

From the Benedictine monk Trithemius, we observe the futility of resisting innovation. Bibliophiles are, perhaps, some of the most egregious offenders. In the same way that Trithemius clung to the power and majesty of handwritten manuscripts, many book lovers today resist the shift to electronic books. Ones and zeros may seem fragile next to paper and cardboard and leather, but Trithemius thought the same thing about the delicate creations of Gutenberg's printing press.

During Mercator's day, vast amounts of information were becoming widely available for the first time. The printing presses of Europe were flooding the continent with accounts of circumnavigations, newly -discovered manuscripts, original treatises, and probably drawings of "Keyboard Cat." Humanity struggled to find a way to process this deluge of information—with all the foibles and dangers that accompanied it—in a way not so different from the advent of websites such as Wikipedia.

From Tyndale's day we see the true power of the words we use. By changing the word *church* to *congregation*, an entire religious sect can be spawned. Pious people can burn their fellow citizens for printing the word *love*, and feel justified for doing it. Words are narrative-defining tools, especially during conflict. In modern times, we needn't look any further than *advanced interrogation techniques* versus *torture*.

As with language, so with history: context matters. It was something William Blake knew all too well as he combined the printed word and the printed illustration into one "composite art." When context is left out, important meanings can get lost in translation. If the scholars who first authenticated De Caro's *Sidereus Nuncius*, for example, had been informed of other known Galileo forgeries, they might have been more primed for doubt and thus have uncovered the counterfeit earlier.

The "bad" plays of Shakespeare challenge the way we perceive the dominant narratives we take for granted. When we put Shakespeare on such a high pedestal, he becomes more statue than human. So it is with his work. The great Bard's plays are richer and more meaningful because he was an actual human. Other actual humans helped bring those plays to the printing presses of the world. They say it takes a village to raise a child, but apparently it also takes a village to create *King Lear.*

The rise and fall and rise of Mary Wollstonecraft makes us question what it means to be an author when people are treated differently not for what they say, but for who they are. Considering that gender equality is not a problem today—sorry, that should read "considering that gender equality is *still* a problem today"—Wollstonecraft's fate is just as relevant in 2017 as it was in 1800.

The network of printing and its influence that spread across the American colonies in Benjamin Franklin's day did so on a river of currency. If you want to trace the corridors of power, you need only follow that money. And when Dickens followed that money right to the doorsteps of American publishers in New York and Massachusetts, the United States promptly said, *Bugger off, Charles Dickens. America steals things.* But before we gasp in horror at the wholesale stealing of another's work, it's worth taking a quick look at how flippantly we post original quotes, photos, designs, and illustrations on the Internet without giving credit.

We might also want to cast a more critical eye on those "nobody reads anymore!" essays online. Recall that in 1921, only 4 percent of Americans visited a bookstore. Book-loving folks who claim everything was better in the old days must have a very limited perception of the word *old*. Many of our assumptions about what book culture "should" be were formed only in the twentieth century, by men such as Eddie Bernays, using clever marketing techniques we don't even notice. "We live in an expanding culture," the historian Raymond Williams has said, "yet we spend much of our energy regretting the fact, rather than seeking to understand its nature and conditions."

T. J. Cobden-Sanderson understood the power of beauty, in his own way. Ironically, his disdain for technological innovation led him to create some of the most innovative books of his time. When he hurled his Doves Type into the Thames, he was expressing a fear that people still harbor: that advancements have the potential to rob books of their essence. But what is a book? Is it a codex? Is it a scroll? A handwritten manuscript? Is a true book the result of a single-pull printing press? Or a double-pull? Or a Linotype machine? Does it have to be physically printed at all? Can the letters on a screen match the power and resonance of wood, cloth, and leather?

We exist in a time that is not as unique as it first appears. Many of the issues we face in the digital age have been faced before. History is a window that lets us look backward for answers. It's a window we should be constantly revisiting. The historian Randolph G. Adams said, "Each generation has to rewrite history for itself—and sometimes from the same sources used by previous generations." How do we resolve disputes over public domain or Net neutrality in 2017? Just ask Mr. Dickens. He has firsthand experience trying to change American minds about copyright. In 1867 he declined even to talk about the issue, "on the grounds that he felt the case to be a hopeless one." Okay then, fine, don't ask Mr. Dickens.

If history is that glorious window to our past, remember that this book is the tailgate party. In the parking lot. Of the building down the street from the house in which said glorious window resides. So pull yourself up from the pavement, wipe away the vomit in your hair, and crack open that biography of Thomas More.

ACKNOWLEDGMENTS

The authors would like to thank the following persons and institutions, without whom this work would never have been possible: Cal Morgan, who believed in us; Michelle Brower, our rockstar agent; Stephanie Hitchcock, our precise and long-suffering editor; Bauman Rare Books and Honey & Wax Booksellers, where Rebecca learned her trade; the specialists and advance readers who gave us notes: Nick Wilding, Russel Maret, Adam Hooks, Rachel D'Agostino, John Windle, Elizabeth Denlinger, Vic Zoschak, Robert Green, Ina Saltz, Loren Glass, Mitch Fraas, Jonathan Kearns, Ben Skerker (to whom credit is due for the Feng Shui joke), and Mike Smith; institutions that were kind to us, especially University of Virginia's Rare Book School and University of Pennsylvania; our children, who are too young to have been anything other than obstructions, but screw it, thanks, Elliott and Anson; and lastly, a special *danke* to Johann Gensfleisch zur Laden zum Gutenberg: you weren't the first to invent movable type, and virtually no one credited you properly for two hundred fifty years, but you gave the world a European single-pull printing press and without you the subjects of this book (and the physical book itself) would not exist. Rest in peace, bro.

NOTES

Introduction: What Do You Reckon This Is?

xi *first edition of the King James Bible*: A book that A. S. W. Rosenbach described four hundred years later as "in every respect one of the finest things a collector can ever hope to acquire." *Books and Bidders: The Adventures of a Bibliophile* (Boston: Little, Brown and Company, 1927), 237.

xii *"Thou shalt commit adultery"*: Henry Richard Tedder, "Barker, Robert," *Dictionary of National Biography*, vol. 3: *Baker–Beadon* (London: Smith, Elder, and Company, 1885), 207–8.

xii *"mistake"*: In quotes because, really, *was* it a mistake? Some scholars suggest it may have been sabotage.

xii *tens of thousands of dollars*: The most recent copy on the market sold for more than $45,000 in November 2015. Bonhams, the auction house, records only five copies appearing at auction since World War II. See https://www.bonhams.com/auctions/22715/lot/5/.

xii *and a penis*: For a full play-by-play, see Kevin Mac Donnell, "Huck Finn Among the Issue Mongers," *FIRSTS* 8, no. 9 (September 1998): 28–35.

xiii *five-hundred-dollar reward*: *New York Herald*, November 29, 1884. Concerning the problem of finding the culprit, Mac Donnell memorably remarks, "Webster was unable to identify the culprit because his pressroom employed 50 people, all with access to the plates. Most, if not all, of those employees had penises." Not to criticize this lovely explanation, but people without penises can draw them, too. Mac Donnell, "Issue-Mongers," 32.

xiii *"greatest American humorist"*: Obituary, *New York Times*, April 22, 1910. http://twain.lib.virginia.edu/sc_as_mt/mtobit8.html.

xiv *ended up displayed*: *New York World*, November 28, 1884.

xiv *"sole cause of consciousness"*: Fyodor Dostoyevsky, *Notes from Underground*, trans. by Richard Pevear and Larissa Volokhonsky (New York: Vintage Books, 1994), 34–35.

xiv *"dead, but they are there"*: Yakov Smirnoff, *America on Six Rubles a Day* (New York: Vintage Books, 1987), 118.

xv *Who gets to determine how we are remembered*: Cf. Lin Manuel Miranda, *Hamilton: An American Musical*: "Who lives, who dies, who tells your story?"

xvii *"cause of madness"*: Richard D. Altick, *The English Common Reader: A Social History of the Mass Reading Public, 1800–1900*, foreword by Jonathan Rose (Columbus: Ohio State University Press, 1957), 37.

xvii *"or you wouldn't have come here"*: Lewis Carroll, *Alice's Adventures in Wonderland*, a facsimile of the original edition (New York: Alfred A. Knopf, 1984), 90.

Chapter 1: How to Forge a Rare Book

1 *presented for sale*: According to his 2013 interview with Nicholas Schmidle, De Caro was working with a partner in offering the book, Filippo Rotundo. See Nicholas Schmidle, "A Very Rare Book: The Mystery Surrounding a Copy of Galileo's Pivotal Treatise," *The New Yorker*, December 16, 2013, http://www.newyorker.com/magazine/2013/12/16/a-very-rare-book.

2 *"acquisition of a lifetime"*: Ibid.

2 *so unheard of*: As Nick Wilding pointed out in conversation with the authors, many types of forgeries have appeared over the centuries. Piracies, a complaint of printers as early as the fifteenth century, are an ever-present example. However, we both agree that the type of forgery here (the production of a complete item from scratch with the latest technology, specifically to be sold on the collectible market) is a new and particularly modern event. The oldest of its kind may be T. J. Wise's early twentieth-century forgeries of poetical pamphlets sold to collectors, which were uncovered in a book with the most boring name possible for the bomb it truly was: John Carter and Graham Pollard, *An Enquiry into the Nature of Certain Nineteenth Century Pamphlets* (1934). See also Nick Wilding, *Faussaire de Lune: Autopsie d'une imposture, Galilée et ses contrefacteurs*, trans. Antoine Coron (Paris: Bibliothèque nationale de France, 2015), 14.

3 *forgeries of Galileo's treatises:* This one was uncovered, ironically, by Owen Gingerich, one of the scholars who initially signed off on the *Sidereus Nuncius*. A couple years later, however, Gingerich registered doubt about the phases of the moon depicted in the watercolors before the authenticity of the entire treatise was suspected. See Owen Gingerich, "The Curious Case of the M-L *Sidereus Nuncius*," *Galilæana* 6 (2009): 141–66; Nick Wilding, review of *A Galileo Forgery: Unmasking the New York* Sidereus Nuncius," Horst Bredekamp, Irene Brückle, and Paul Needham, eds., in *Renaissance Quarterly* 67, no. 4 (2014): 1340; and Horst Bredekamp et al., "Introduction," in *A Galileo Forgery*, ed. by Horst Bredekamp et al. (Berlin: De Gruyter, 2014), 10.

3 *word* masterpiece *was thrown around*: Irene Brückle, "Final Thoughts," in Bredekamp et al., *A Galileo Forgery*, 100.

3 *embarrassed that it took so long*: Paul Needham said, "But is it a clever forgery? I am not convinced, despite being someone who managed to be fooled by both its printing and its paper. I am happier in saying this reflects poorly on me than that it reflects well on the makers." Paul Needham, "Final Thoughts," in Bredekamp et al., *A Galileo Forgery*, 95.

4 *through 4,200 pages of all Galileo's known letters*: Schmidle, "A Very Rare Book."

4 *"an extended undergraduate paper with no quotations"*: Quoted in ibid., pulled off *The New Yorker* website.

5 *after De Caro's crimes were uncovered*: Ibid.

5 *replaced these authentic editions with his forgeries*: Wilding, *Faussaire*, 47.

6 *"so hot right now"*: See *Zoolander* (2001), DVD. The actual quote is "Hansel, so hot right now. Hansel."

6 *studied papermaking by hand in Italy*: Schmidle, "A Very Rare Book."

6 *it's quite the opposite*: Relatively speaking. Of course before paper was brought to the West—having been invented in China and introduced to Spain and Italy via trading with Muslim communities—animal-based parchment was used for books, which was tremendously more expensive.

7 *equal the cost of everything else combined*: Cristina Dondi, "The European Printing Revolution," in *The Book: A Global History*, ed. Michael F. Suarez, S.J., and H. R. Woudhuysen (Oxford: Oxford University Press, 2013), 87.

7 *"rag sermons"*: Adrian Johns, *The Nature of the Book: Print and Knowledge in the Making* (Chicago, IL: University of Chicago Press, 1998), 17. With thanks to Nick Wilding for pointing out this example.

7 *making their living sifting through refuse piles*: See, for example, Henry Mayhew, *London and the London Poor*, 2 vols. (London: Griffin, Bohn and Company, 1861), 2:138–42.

9 *aromatic organic polymer called lignin*: John Christopher Roberts, *The Chemistry of Paper* (London: Royal Society of Chemistry, 1996), 26.

10 *cotton scraps . . . anachronistic red flags*: Irene Brückle, Theresa Smith, and Manfred Mayer, "The Evidence of the Forged Paper," in Bredekamp et al., *A Galileo Forgery*, 38.

10 *spell in a kitchen oven*: Schmidle, "A Very Rare Book."

10 *faux aging . . . invisible fingerprints*: See Brückle, Smith, and Mayer, "The Evidence of the Forged Paper," in Bredekamp et al., *A Galileo Forgery*, 52.

11 *add lasting pigment to inks*: Joe Nickell, *Pen, Ink, and Evidence* (New Castle, DE: Oak Knoll Press, 2003), 40.

11 *female gall wasp*: Ibid., 36.

11 *his secret formula*: Janet Ing, *Johann Gutenberg and His Bible* (New York: Typophiles, 1988), 87.

12 *Soaking it in pee*: Margreta De Grazia and Peter Stallybrass, "The Materiality of the Shakespearean Text," in *Shakespeare and the Editorial Tradition*, ed. Stephen Orgel and Sean Keilen (London: Taylor and Francis, 1999), 28.

12 *Gutenberg's most important innovation*: Lucien Febvre and Henri-Jean Martin, *The Coming of the Book*, trans. David Gerard (London: Verso, 2010), 50.

12 *you tend to avoid slicing off pieces*: Brückle, Smith, and Mayer, "Evidence of the Forged Paper," 38.

13 *tested acceptably high on acid*: Schmidle, "A Very Rare Book."

14 *unequivocal evidence proving the fraud*: See Nicholas Pickwoad, "The Evidence of the Forged *SNML* Sammelband Book Structure," in Bredekamp et al., *A Galileo Forgery*, 61–70.

14 *expensive slats of wood*: Barbara A. Shailor, *The Medieval Book* (Toronto: University of Toronto Press, 1994), 56.

15 *morocco, made from goatskin*: Modern bookbinders reading this book have now just tossed it aside in disgust. The academic and collecting world of rare books still uses the term "morocco" as a catchall for any goatskin binding, but bookbinders know it's in fact only one specific type of goatskin. With thanks for Lang Ingalls's comments at the Colorado Antiquarian Book Seminar in July 2016, for pointing this out.

16 *bound in Allen's skin*: Nicholas Basbanes, *A Gentle Madness* (New York: Henry Holt and Company, 1999), 154.

17 *cut, gilded, and gauffered at the same time*: Pickwoad, "Book Structure," 70.

17 *fornicating on the fore-edge*: This very copy of Longfellow sold at Heritage Auctions in April 2014 for $406.25.

18 *most successful book forgery*: As Rebecca's Medieval Latin professor, Richard Lounsbury, used to say, does a "successful forgery" really exist? If it is successful, no one calls it a forgery. If the forgery is revealed, then it clearly isn't successful.

18 *faulty capital* P: The first published statement declaring the forgery focused on this detail, a misshapen *P* that could have occurred only through digital retouching. See Nick Wilding, "Letter to the Editor," *ISIS* 103, no. 4 (2012): 760.

18 *uncover his masterful deceit*: Schmidle, "A Very Rare Book."

19 *"a projected duel"*: Ibid.

19 *photopolymer plates*: See Wilding, *Faussaire*, 37–39. This is one of the key differences in modern forgeries, a technological development that has made them much more financially practical. Thomas J. Wise's famous forgeries were, for instance, possible mainly because of his connection with a local print shop, where he had already been printing "authorized" facsimiles of rare first editions.

19 *"real expert" happened to be working*: Nick Wilding, "Review of *Galileo's O*, vols. 1 and 2, edited by Horst Bredekamp," *Renaissance Quarterly* 65, no. 1 (2012): 217–18.

19 *senior scholars blew off Wilding's concerns*: Wilding, "Reviewed Work," 1338.

19 *the key word here being* fantasy: See Needham, "Final Thoughts," 98.

20 *"within about twenty minutes"*: Paul Needham, "Fruitful Doubts, May–June 2012," in *A Galileo Forgery*, 23.

20 *"not many results are worse than a refuted authentication"*: Bredekamp et al., "Introduction," 11.

20 *only a masterpiece could have fooled them*: Wilding, "Reviewed Work," 1338.

20 *seven years' house arrest*: Technically he was sentenced to seven years in prison, which was commuted to house arrest because of his health. "Biblioteca dei Girolamini, condannato l'ex direttore De Caro a 7 anni," *Napoli Today*, March 15, 2013, at http://www.napolitoday.it/cronaca/biblioteca-dei-girolamini-direttore-condannato.html.

20 *being a thief*: He was convicted for the theft of thousands of books from a library in Naples. Even considering the millions of dollars involved, the Italian government doesn't seem to know how to handle these bibliocrimes, which are truly crimes against our intellectual heritage. De Caro still hasn't been brought to trial for the sale of his forged *Sidereus Nuncius*.

Chapter 2: Forgetting Mr. Gooseflesh

21 *"must not be stopped because of printing"*: Quoted in Noël L. Brann, *The Abbot Trithemius (1462–1516): The Renaissance of Monastic Humanism* (Leiden: Brill, 1981), 156. We've used Brann's translations throughout when quoting Trithemius.

21 *differences between handwriting and printing*: Ibid., 156.

22 *Gutenberg appears in exactly zero printed books*: Ing, *Gutenberg*, 33.

22 *birthday, or birth year, or birth decade*: Ibid., 27.

23 *common medieval derivations of "Johannes"*: Douglas C. McMurtrie, *The Gutenberg Documents* (New York: Oxford University Press, 1941), 42.

23 *dually libelous opinion of Gutenberg*: Ing, *Gutenberg*, 48–49. The theory had been tossed out by the eighteenth century in England, but Americans love conspiracies. They kept the theory alive into the nineteenth century. They were also, incidentally, first responsible for putting forth theories questioning William Shakespeare of Stratford-upon-Avon as the true author of the plays attributed to him.

24 *Gutenberg was remembered only as their "assistant"*: Martyn Lyons, *Reading Culture and Writing Practices in Nineteenth-Century France* (Toronto: University of Toronto Press, 2008), 92–94.

24 *didn't appear until 1472*: Ing, *Gutenberg*, 33.

24 *rescue Gutenberg's reputation*: Ibid., 45.

24 *went missing for more than a hundred fifty years*: McMurtrie, *Gutenberg Documents*, 176. It also happened to be the most important document, the Helmasperger Instrument. More on this record to follow.

25 *"the abuses of a stepfather"*: Brann, *Abbot Trithemius*, 3.

26 *"grant you whatever you have asked"*: Ibid., 5.

27 *"'proud and temperamental' character"*: Paul Needham, *The Invention and Early Spread of European Printing* (Princeton, NJ: Princeton University Library, 2007), 5.

27 "GRAUNT—*popular poet of ancient times*": Mark Twain, *The Innocents Abroad; Or, the New Pilgrims' Progress* (Hartford, CT: American Publishing Company, 1869), 336.

28 *zombies hanging around being jerks*: Okay, yes, there is a reason: the woodcut forms part of a larger *Danse Macabre* theme, a popular medieval trope that emphasized the ubiquity of death and how it comes for people in every station in life. The *Walking Dead* of the Middle Ages, if you will.

28 *"we know nearly nothing"*: Ing, *Gutenberg*, 78.

30 *deal with Satan*: Elizabeth L. Eisenstein, *Divine Art, Infernal Machine: The Reception of Printing in the West from First Impressions to the Sense of an Ending* (Philadelphia: University of Pennsylvania Press, 2011), 2–6.

30 *Gutenberg kept printing*: Paul Needham, "Johann Gutenberg and the Catholicon Press," *The Papers of the Bibliographic Society of America* 76, no. 4 (1982): 432.

30 *Gutenberg has left after his death*: McMurtrie, *Gutenberg Documents*, 219.

30 *suit may actually have favored Gutenberg*: Ing, *Gutenberg*, 31.

30 *Gutenberg was cast as a nosy neighbor*: Ibid., 43.

31 *previously unknown eyewitness account*: Ibid., 67.

31 *"earliest precise date by which we know typographic printing"*: Needham, *Invention and Early Spread*, 17.

31 *a note nonchalantly scribbled*: Ing, *Gutenberg*, 53.

31 *literary casualty list*: See McMurtrie, *Gutenberg Documents.*

32 *liveliness to times and circumstances*: Quoted in Brann, *Abbot Trithemius*, 151.

32 *marvelous and hitherto unheard of art of printing*: Quoted in ibid., 145.

32 *O blessed art of printing*: Quoted in ibid., 147.

33 *"Sponheim Abbey Press"*: Ibid., 149.

34 *"I will gladly go to labors outside"*: Quoted in ibid., 165.

34 *"writer who commends his writings to membranes"*: Quoted in ibid., 151.

34 *reap the following rewards*: Quoted in ibid., 155.

35 *"will be able to endure a thousand years"*: Quoted in ibid., 157.

35 *rate of nearly two to one*: Bettina Wagner, ed., *Als die Lettern laufen lernten. Medienwandel im 15. Jahrhundert* (Wiesbaden: Reichert, 2009), 15.

35 *Amazon erased illegal editions of George Orwell's*: Brad Stone, "Amazon Erases Orwell Books from Kindle," *New York Times*, June 17, 2009.

36 *circulation in manuscript form*: See Richard B. Wollman, "The 'Press and the Fire': Print and Manuscript Culture in Donne's Circle," *Studies in English Literature, 1500–1900* 33, no. 1 (Winter 1993): 85–97.

36 *"distinguishing their beautiful volumes"*: Brann, *Abbot Trithemius*, 158.

37 *"a favorite, old blanket"*: David Scott Kastan, *Shakespeare and the Book* (Cambridge, UK: Cambridge University Press, 2001), 113.

37 *papyrus with processed animal carcasses*: An Egyptian would also have turned his nose up at the idea of leather competing with papyrus at all—the word *parchment* is rooted in the proper noun Pergamum, the ancient Greek city that produced leather for books specifically in order to compete with Egypt's papyrus production.

38 *"highly favorable one"*: Brann, *Abbot Trithemius*, 145.

38 *"library of two-thousand books"*: Ibid., 53.

38 *"abounding with volumes"*: Ibid., 70–71.

38 *different can of* kerscheblotzer: A cherry cake popular in the Mainz area. Not served in a can—as far as we know.

39 *"indeed without glasses"*: Quoted in Martin Davies, "Juan de Carvajal and Early Printing: The 42-line Bible and Sweynheym and Pannartz Aquinas," *The Library* (1996).

39 *30 percent of editions from this period*: Jonathan Green and Frank McIntyre, "Lost Incunable Editions: Closing in on an Estimate" in *Lost Books: Reconstructing the Print World of Pre-Industrial Europe*, ed. Flavia Bruni and Andrew Pettegree (Leiden: Brill, 2016), 60–61. This article also discusses the value and drawbacks of using ISTC to count incunable editions.

39 *Latin term coined in 1640*: Bernhard von Mallinckrodt suggested the term in *De ortu et progressu artis typographicae*. Cf. Needham, *Invention and Early Spread*, 9.

39 *babies of print*: Does this make the pamphlet wars of the Reformation its terrible twos? Cf. chapter 3.

40 *"are to be spat at"*: Grant Uden, *Understanding Book-Collecting* (New York: Antique Collectors' Club, 1982), 36.

40 *probably trained in Mainz*: Christopher de Hamel, *The Book: A History of the Bible* (London: Phaidon Press, 2005), 192.

40 *excellent candidates for the title of first printed book*: Paul Needham, "Precious Consignments from the Old to the New World: The Gutenberg Bible in America," in *Association internationale de bibliophilie: Actes et Communications*, XXVth Congress, New York City, and Post-Congress, Chicago (s.l.: Association international de bibliophilie, 2011), 46.

41 *copyediting errors while using B42*: Karl Dziatzko, *Gutenbergs früheste Druckerpraxis* (Berlin: A. Asher and Co., 1890). See Ing, *Gutenberg*, 54–57, for a summary of Dziatzko's findings.

42 *"purchased without great sacrifice"*: Ibid., 105, summarizing the conclusions of Eberhard König.

42 *170 calves or 300 sheep*: Sarah E. Bond, "Sacrificial Lambs: Livestock, Book Costs, and the Premodern Parchment Trade," *Sarah E. Bond: Late Antiquity, Digital Humanities, and Musings on the Classical World*, April 2016, https://sarahemilybond.wordpress.com/2016/04/04/sacrificial-lambs-livestock-book-costs-and-the-premodern-parchment-trade/.

42 *largest herd of cattle in the United States*: Colleen Schreiber, "Mormon Church Holds Title as Largest U.S. Cow-Calf Producer," *Livestock Weekly*, January 25, 2001, http://www.livestockweekly.com/papers/01/01/25/whldeseret.asp.

43 *started placing bids*: Technically, this is also when Sir Thomas started placing bids at all; up to this point he had been sitting back like a gentleman, preferring his agent to bid. A man obviously paid for his cool tempera-

ment: this agent ceased at Sir Thomas's planned bid. That's when Sir Thomas stepped in himself.

43 *"arrest [Sir Thomas's] mad career"*: Henry Stevens, *Recollections of James Lenox of New York and the Formation of His Library* (London: H. Stevens and Son, 1886), 32.

43 *"swallowed up by the Deep Sea"*: Quoted in Ezra Greenspan, *George Palmer Putnam: Representative American Publisher* (University Park, PA: Penn State Press, 2010), 158.

44 *denied the possession of these Bibles*: Needham, "Precious Consignments," 40.

44 *"These people were not art specialists"*: "Russia Sentences Secret Agents over Theft of Gutenberg Bible," *BBC News*, June 6, 2014, http://www.bbc.com/news/world-europe-27738164.

45 *where Gutenberg's print shop stood*: Ing, *Gutenberg*, 75.

45 *running two separate print shops*: Ibid.

45 *found inside the binding of an accountant's book*: Colin Clair, *A History of European Printing* (London: Academic Press, 1976), 16.

45 *church where he was likely buried was demolished*: Seán Jennett, *Pioneers in Printing* (London: Routledge and Kegan Paul Ltd., 1958), 21.

46 *monastic issues, histories, and demonology*: Please thank the Oxford comma for keeping this phrase from reading, "monastic issues, histories and demonology."

47 *taught him the ancient secrets of steganography*: Noël L. Brann, *Trithemius and Magical Theology* (New York: State University of New York Press, 1999), 101.

48 *"offend them by various delusions"*: Ibid., 136.

49 *branded it a work of heresy*: Ibid., 172.

49 *demonology, "a minor current"*: Brann, *Abbot Trithemius*, 101.

Chapter 3: Trees of Truth

52 *surrender his dead son's sheet*: Susan Brigden, *London and the Reformation* (London: Faber and Faber, 2014), 99.

52 *"thou art accursed"*: Quoted in ibid.

52 *"Wycliffe's damnable works"*: J. Fines, "A Post-Mortem Condemnation for Heresy of Richard Hunne," *English Historical Review* 78, no. 308 (July 1963): 530.

53 *"plenty of blood was shed before he was hanged"*: John Foxe, *The Acts and Monuments of John Foxe: With a Preliminary Dissertation by the Rev. George*

Townsend, 8 vols., ed. Rev. Stephen Reed Cattley (London: R. B. Seeley and W. Burnside, 1837–1841), 4:191.

54 *"the soul from purgatory springs"*: This popular verse is referenced directly in statement number 27 of Luther's Ninety-Five Theses.

54 *splinters from Christ's True Cross to fill a ship*: Erasmus, *Colloquies*, vol. 2, trans. N. Bailey (London: Reeves and Turner, 1878), 13.

54 *gnawed his way through them with his molars*: Jonathan Sumption, *The Age of Pilgrimage: The Medieval Journey to God* (Mahwah, NJ: Paulist Press, 2003), 41.

54 *"a few of my neighbors"*: Quoted in Preserved Smith, *The Life and Letters of Martin Luther* (Boston: Houghton Mifflin, 1911), 43.

55 *"printed and circulated far beyond my expectation"*: Quoted in ibid., 44.

55 *"had I known what was going to happen"*: Quoted in ibid.

55 *printing in the common German tongue*: Andrew Pettegree, *Reformation and the Culture of Persuasion* (Cambridge, UK: Cambridge University Press, 2005), 164.

56 *"Heralded on all sides as a 'peaceful art'"*: Elizabeth Eisenstein, *The Printing Press as an Agent of Change* (Cambridge, UK: Cambridge University Press, 1979), 319.

56 *could that mule have been crucified*: Brian Moynahan, *God's Bestseller* (New York: St. Martin's Press, 2014), 10.

56 *what was actually written in the Bible*: S. L. Greenslade, "Epilogue," in *Cambridge History of the Bible*, ed. Greenslade (Cambridge, UK: Cambridge University Press, 1976), 485.

56 *"did not know how many Commandments there were"*: W. M. S. West, "John Hooper and the Origins of Puritanism" (Zurich dissertation, 1955), 45.

57 *"a boy that driveth the plough"*: Quoted in David Daniell, *William Tyndale: A Biography* (New Haven, CT: Yale University Press, 1994), 1.

57 *"Arthur Cobbler" or "Hans Hoe"*: A vicar called the Bible in English "the book of Arthur Cobbler," as quoted in G. R. Elton, *Policy and Police: The Enforcement of the Reformation in the Age of Thomas Cromwell* (Cambridge, UK: Cambridge University Press, 1972), 27; "Hans Hoe" was the German nickname, a translation of "Karsthans" in German: Mark U. Edwards Jr., *Printing, Propaganda, and Martin Luther* (Minneapolis, MN: Fortress Press, 2005), 61.

57 *crime punishable by death*: This was the end result of laws that combined the judgments of spiritual courts with the enforcement by governmental courts: Parliament's 1401 passage of *De heretico comburendo* (allowing burning at the stake as lawful punishment for heretics) and Archbishop Arun-

del's 1407–9 passage of Constitutions, or religious regulations, one of which forbade the ownership or circulation of any written translation of scripture into English. See Nicholas Watson, "Censorship and Cultural Change in Late-Medieval England: Vernacular Theology, the Oxford Translation Debate, and Arundel's Constitutions of 1409," *Speculum* 70, no. 4 (October 1995): 822–64.

57 *"not the light of the world, but its darkness"*: Quoted in Moynahan, *God's Bestseller*, 156.

57 *"indeed women and simple idiots"*: Johannes Cochlaeus, quoted in ibid., 33.

57 *"cartload of hay for a few sheets of St. Paul"*: Ibid., xix.

58 *sold his looms and shears to purchase a copy*: Brigden, *London*, 89.

58 *"advised me to seek in London"*: Quoted in Daniell, *Tyndale*, 85.

59 *encouraged printers to issue more pirated editions*: J. F. Mozley, *William Tyndale* (New York: Macmillan, 1937), 115–19.

59 *vernacular translations of the Bible were widely available*: John L. Flood, "The History of the Book in Germany," in *The Book: A Global History*, ed. Suarez and Woudhuysen (Oxford: Oxford University Press, 2013), 370.

59 *"vagabond or religious eccentric"*: Anne Hudson, ed., *Selections from the English Wycliffite Writings* (Cambridge, UK: Cambridge University Press, 1981), 8.

59 *"Ten Worst Britons" poll*: "'Worst' Historical Britons Named," *BBC News*, December 27, 2005, http://news.bbc.co.uk/2/hi/uk/4560716.stm.

59 *legally enforce these spiritual rulings*: Legal enforcement of spiritual courts against heretics began in 1401, with the passing of *De heretico comburendo*. See note on page 57.

60 *sixty manuscript copies of Chaucer's English masterpiece*: De Hamel, *The Book*, 187.

60 *heresy . . . with sedition*: Richard Rex, "Thomas More and the Heretics: Statesman or Fanatic?" in *The Cambridge Companion to Thomas More*, ed. George M. Logan (Cambridge, UK: Cambridge University Press, 2011), 95.

61 *their beloved English New Testaments tied around their necks*: De Hamel, *The Book*, 186.

61 *"to the end without moving"*: Foxe, *Acts and Monuments*, 4:688.

61 *"not God's body"*: Ibid., 3:238.

61 *capitalist innovators and Lollard heretics*: E. G. Rupp, *Studies in the Making of the English Protestant Tradition* (Cambridge, UK: Cambridge University Press, 1966), 8.

62 *bales of cloth*: David Daniell, "William Tyndale, the English Bible, and the

English Language," in *The Bible as Book*, Vol. 3: *The Reformation*, ed. Orlaith O'Sullivan (New Castle, DE: Oak Knoll Press, 2000), 43. Thomas More also learned the trick with the bales of cloth—through the interrogation of George Constantine, a colporteur known as a major distributor of English New Testaments. Some scholars have suggested that Constantine first began circulating rumors about extreme treatment of prisoners by More in order to account for the large amount of information Constantine betrayed during his questioning. See Brad C. Pardue, *Printing, Power, and Piety: Appeals to the Public During the Early Years of the English Reformation* (Boston: Brill, 2012), 108–9. For the works Tyndale used in his translation, see Daniell, *Tyndale*, 108–15.

63 *"in short time be Lutheran"*: Quoted in Daniell, *Tyndale*, 109.

63 *"devil in the habit of a monk"*: The Edict of Worms, 1521.

64 *wasn't identified until 1996*: See Eberhard Zwink, "Confusion About Tyndale: The Stuttgart Copy of the 1526 New Testament in English," *Reformation* 3, no. 1 (1998).

64 *purchased for six days' wages*: Pardue, *Printing*, 19.

64 *"cost him just 20 guineas"*: "Tyndale New Testament," British Library, http://www.bl.uk/onlinegallery/sacredtexts/tyndale.html.

65 *would not sit for portraits*: Moynahan, *God's Bestseller*, 54.

65 *"hellhound in the kennel of the devil"*: Quoted in Thomas More, *Confutation of Tyndale's Answer*, vol. 8 of *The Complete Works of St. Thomas More*, ed. Louis A. Schuster, Richard C. Marius, and James P. Lusardi (New Haven, CT: Yale University Press, 1973), 226, 135, 358. Thomas More, *A Dialogue Concerning Heresies*, vol. 6 of *The Complete Works of St. Thomas More*, ed. Thomas M. Lawler, Germain Marc'hadour, and Richard C. Marius (New Haven, CT: Yale University Press, 1981), 424.

65 *all the powers of Church and State*: Though a layman, More was commissioned by the bishop of London, Cuthbert Tunstall, to fight against heresy with publications in English. See Eamon Duffy, "'The Comen Knowen Multitude of Crysten Men': *A Dialogue Concerning Heresies* and the Defence of Christendom," in Logan, ed., *Cambridge Companion to Thomas More*, 194.

65 *"the simple and unlearned"*: Quoted in Daniell, *Tyndale*, 69.

66 *Peasants' War cited Luther's rebellion*: More argues that the Peasants' War was an "inevitable outcome" of Luther, noted in Duffy, "Defence of Christendom," 209.

66 *"rebels as well as heretics"*: Bridgen, *London*, 120.

67 *to some extent authored*: For more on More's contribution to Henry VIII's pamphlet, see Peter Ackroyd, *Life of Thomas More* (New York: Nan A. Talese, 1998), 226.

67 *"covered in excrement"*: Quoted in ibid., 227.

67 *"shit, dung, filth, and excrement"*: Quoted in ibid., 230–31. Ackroyd does a lovely job of providing both the Latin and the English terms for any readers who would like to expand their Latin vocabulary.

67 *"the greatest heap of nasty language"*: Ibid., 227, quoting Francis Atterbury. These references feature in a chapter of Ackroyd's book called "XXI: I Am Like a Ripe Shit"—named for a quote from More, naturally.

67 *"obsessive anality"*: Daniell, *Tyndale*, 277. As a counterpoint, note Richard Rex's criticism of characterizing More's writings based on these passages: "His *Responsio ad Lutherum* delves into the dungheap on perhaps a dozen occasions, two or three of them spectacularly revolting. But this is not typical of a text which amounts to about 350 pages in its critical edition." Only a dozen occasions! Rex, "Statesman or Fanatic," 104–5.

68 *accurate and faithful to the original texts*: Moynahan, *God's Bestseller*, 104. For a discussion of what the committee of scholars making the King James translation owed (and didn't owe) to Tyndale, see Adam Nicolson, *God's Secretaries: The Making of the King James Bible* (Grand Rapids, MI: Zondervan, 2005), 221–23.

68 *came down to three words*: More, *Dialogue*, 285–86. See also Richard Duerden, "Equivalence or Power? Authority and Reformation Bible Translation," in O'Sullivan, ed., *The Bible as Book: The Reformation*, 12.

70 *international mecca for printing*: Dr. F. de Nave, "Antwerp, Dissident Typographical Centre in the Sixteenth Century: General Synthesis," in *Antwerp, Dissident Typographical Centre*, ed. Gilbert Tournoy, Dirk Imhof, and Francine de Nave (Antwerp: Snoeck-Ducaju and Zoon, 1994), 11.

70 *"farthest out of danger"*: Rev. R. Demaus, *William Tyndale, a Biography*, rev. Richard Lovett (London: Religious Tract Society, 1886), 319

70 *surpassing the pope*: In Tyndale's *Obedience of a Christian Man* (Antwerp: Merten de Keyser, 1528).

70 *unusually competent ruler*: See David Starkey, *Six Wives: The Queens of Henry VIII* (New York: HarperCollins, 2003).

70 *"failure in the home"*: David O. McKay, in *Family Home Evening Manual*, ed. the Council of the Twelve Apostles (Salt Lake City, UT: The Council, 1968), iii.

71 *transmute the religious landscape of England*: To complicate matters, in 1527 Catherine's nephew Charles V had recently sacked Rome, reducing the pope to a mere puppet during a crucial period of the "Great Matter."

71 *she was a fan*: For an examination of the sources describing Anne Boleyn's reformer leanings, see Thomas S. Freeman, "Research, Rumour and Propaganda: Anne Boleyn in Foxe's 'Book of Martyrs,'" *The Historical Journal* 38 no. 4 (December 1995): 797–819.

71 *connections in the smuggling market*: Bridgen, *London*, 128.

71 *"servants were all encouraged to read"*: Reported by Boleyn's confessor, Latimer, quoted in Moynahan, *God's Bestseller*, 295.

71 *presented Henry with her favorite passages*: Ackroyd, *Thomas More*, 283, referencing Foxe.

72 *allow an English translation*: Daniell, *Tyndale*, 212.

72 *"I did but warn his grace"*: Quoted in ibid., 213.

73 *"Forbidden Book of the Month Club"*: See Patrick Collinson, "William Tyndale and the Course of the English Reformation," in *Reformation* 1, no. 1 (1996): 72–97.

73 *lurked More's network*: Ackroyd, *Thomas More*, 300.

73 *Facing rumors of torture and indefinite confinement*: Moynahan, *God's Bestseller*, 208. While there is ample documentation for the sustained imprisonment of suspects—John Petyt died while imprisoned in the Tower of London, even though they had no incriminating evidence of his crimes—it's worth noting that More asserted that he never tortured his prisoners. Personally, given how seriously he took the threat of religious fanatics, we're inclined to believe it's because he considered his "advanced interrogation techniques" necessary for the safety of the state.

73 *burn* everyone *he detained*: In the two and a half years of his chancellorship, six heretics were burned. Of these, the cases of three were pressed through specifically by More. See Rex, "Statesman or Fanatic," 94.

74 *"Tree of Truth"*: See Foxe, *Acts and Monuments*, 4:698. Apologists claim, understandably, that Foxe is not the most impartial source for More. Yet we do know that the rumors were common enough that More himself felt obliged to address them.

74 *"bodily harm done him or foul word spoken"*: Thomas More, *Apology*, in *English Prose*, Vol. 1: *Fourteenth to Sixteenth Century*, ed. Henry Craik (1916).

76 *"liberation of language itself"*: David Daniell, "No Tyndale, No Shakespeare," paper presented at the Tyndale Kirtling Meeting, Suffolk, April 16, 2005.

76 *path blazed by Tyndale*: Pardue, *Printing*, 209.

76 *came from his mind*: Daniell, *Tyndale*, 1.

76 *Inquisition "in the Spanish manner"*: Prof. Dr. G. Latré, "William Tyndale in Antwerp: Reformer, Bible Translator, and Maker of the English Language," in *Antwerp, Dissident Typographical Centre*, ed. by Gilbert Tournoy, Dirk Imhof, and Francine de Nave (Antwerp: Snoeck-Ducaju and Zoon, 1994), 64.

77 *the man who doomed Tyndale*: Daniell, *Tyndale*, 368–69.

78 *closely matches Thomas More's profile*: Richard Rex scorns the idea that More could have orchestrated the arrest from his prison in the Tower. He does acknowledge that the targets of this new Protestant movement were "More's kind of people," the educated upper classes. See Rex, "Statesman or Fanatic," 111.

78 *"consumed at last with lice"*: Foxe, *Acts and Monuments*, 5:129.

78 *petitions caused delays*: Daniell, *Tyndale*, 376.

78 *"above all other powers"*: Miles Coverdale, preface to *The Bible, That Is the Holy Scripture of the Old and New Testament, Faithfully Translated into English* (Antwerp: Merten de Keyser, 1535), sig. +ii^{v}.

79 *the name of a heretic*: David Daniell, *The Bible in English* (New Haven, CT: Yale University Press, 2003), 193.

79 *Tyndale's translation would account for*: Jon Nielson and Royal Skousen, "How Much of the King James Bible Is William Tyndale's?," in *Reformation* 3, no. 1 (1998): 49–74.

80 *"forgotten ghost of the English language"*: Collinson, "William Tyndale," quoting from Philip Howard, "Philip Howard Column: Tyndale's Language of the Common Man Is the Bedrock of English Literature Today," in *The Times*, April 29, 1994.

Chapter 4: Making the Round World Flat

83 *sent to the Arctic by none other than King Arthur*: For a full narration of this episode see Thomas Green, "John Dee, King Arthur, and the Conquest of the Arctic," *The Heroic Age: A Journal of Early Medieval Northwestern Europe* 15 (October 2012).

83 *"put into writing all the wonders"*: Letter from April 20, 1577, from Mercator to Dee, quoted in Nicholas Crane, *Mercator: The Man Who Mapped the Planet* (New York: Henry Holt and Company, 2014), 242–44.

84 *legally owns the moon*: We don't actually own the moon. Cf. the 1967 Outer Space Treaty, which incidentally reads almost like a *Star Trek* script.

84 *"pygmies whose length in all"*: The description on the map itself reads, "Pygmei hic habitant / 4 ad summum pedes longi."

84 *"most salubrious of the whole of Septentrion"*: *Hic insula optima est et saluberrima totus Septentrionis.*

85 *"discovered than in five thousand years before"*: Andrew Taylor, *The World of Gerard Mercator* (New York: Walker and Company, 2004), 36.

85 *"info-lust"*: Ann M. Blair, *Too Much to Know: Managing Scholarly Information before the Modern Age* (New Haven, CT: Yale University Press, 2010), 6.

85 *"swarms of new books"*: Seneca and Erasmus both quoted in ibid., 15 and 55.

87 *Columbus carried a copy of the* Geographia: Taylor, *World of Gerard Mercator*, 14.

87 *Jerusalem . . . lay at its center*: Norman J. Thrower, *Maps and Civilization: Cartography in Culture and Society* (Chicago, IL: University of Chicago Press, 2008), 42.

88 *"maps of Scriptural dogma"*: Daniel J. Boorstin, *The Discoverers* (New York: Vintage Books, 1985), 101.

88 *groats per* viertel: Crane, *Mercator*, 14.

88 *strictly forbidden*: In some areas, dissections were allowed, but only infrequently and under extremely strict conditions. During this same period the famous anatomist Andreas Vesalius was forced to grave-rob and scavenge for bodies to autopsy in service of his revolutionary pedagogical anatomy book *De humani corporis fabrica* (1543).

88 *booksellers who went from town to town*: Ibid., 34.

89 *"began to doubt the truth of all philosophers"*: Our translation. Quoted in Latin in Jean Van Raemdonck, *Gérard Mercator, sa Vie et ses Oeuvres* (St. Nicholas, Belgium: Dalschaert-Praet, 1869), 25.

89 *"Let it be said in a whisper"*: Quoted in J. H. Parry, ed., *The European Reconnaissance* (London: Macmillan, 1968), 180.

89 *cartographic revolution was a reaction*: Thrower, *Maps and Civilization*, 58.

90 *shaping that morass*: "Printing diffused more broadly than ever before existing techniques for managing information and encouraged experimentation with new ones, including new layouts, finding devices, and methods of composition." Blair, *Too Much to Know*, 14.

90 *alphabetization*: This section relies on Peter Burke, *A Social History of Knowledge: From Gutenberg to Diderot* (Cambridge, UK: Polity Press, 2000), 184–87.

91 *After print*: "Print" referring, in this case, to the development of illustration techniques used alongside or in conjunction with movable type: relief (woodcut) and intaglio (engraving).

91 *"a globe of exquisite beauty"*: Crane, *Mercator*, 67.

92 *"frivolous distraction"*: Ibid., 61.

93 *Taprobane, the world's largest fake island*: Ibid., 72. It should be noted that this fake island is different from the other fake island required by Ptolemy, and a mainstay of maps for centuries to come: the Great Southern Continent, not definitively disproved until the voyages of Captain Cook in the late eighteenth century.

94 *reduced to a mere abbreviation*: Taylor, *World*, 204.

94 *Mercator wall paneling*: Although it would almost certainly have to come from a map originally issued in a book—maps produced for hanging on walls tend not to survive because they are exposed to the elements more than those hidden between leather covers.

94 *"instruments of rule" . . ."science of princes"*: Michael Biggs, "Putting the State on the Map: Cartography, Territory and European State Formation," *Comparative Studies in Society and History* 41, no. 2 (1999): 380; and J. B. Harley, "Silences and Secrecy: The Hidden Agenda of Cartography in Early Modern Europe," *Imago Mundi* 40 (1988): 59.

94 *punishable by death*: Harley, "Silences and Secrecy," 61.

94 *"secrets not fit to be published"*: Quoted in Roger M. McCoy, *On the Edge: Mapping North America's Coasts* (Oxford: Oxford University Press, 2012), 74.

95 *"knowing is half the battle"*: "GI Joe: A Real American Hero," animated TV series (1985–86). PSAs can be accessed at Internet Archive: https://archive.org/details/GI_Joe-Knowing_is_Half_the_Battle_PSAs.

95 *were classified as state secrets*: Mark Monmonier, *How to Lie with Maps* (Chicago, IL: University of Chicago Press, 1996), 113.

95 *Africa was treated as a state secret*: Burke, *Social History*, 143.

95 *Roman architecture was considered too dangerous*: Ibid., 144.

96 *ensure demons couldn't hide in his hair*: Crane, *Mercator*, 104.

97 *triangulation . . . back-pedaling . . . imperial ass-kissing*: Ibid., 105–8.

98 *"suspect letters"*: Ibid., 136.

99 *"constructing future transgression"*: John H. Arnold, *Inquisition and Power: Catharism and the Confessing Subject in Medieval Languedoc* (Philadelphia: University of Pennsylvania Press, 2001), 85.

100 *arrested at the same time*: Taylor, *World*, 115.

100 *"most unjust persecution"*: Letter of Mercator to Antoine Perrenot de Granvelle, October 9, 1544, quoted in Taylor, *World*, 116.

101 *whole mess of problems*: For more on the technical challenges of representing a sphere on a flat surface, see John P. Snyder, *Flattening the Earth: Two Thousand Years of Map Projections* (Chicago, IL: University of Chicago Press, 1993), 1–4.

101 *not commonly known before Mercator*: Snyder notes that "Erhardt Etzlaub (c. 1460–1532) of Nuremberg used similar projections for a small map limited to Europe and North Africa" in 1511, "but the principle remained obscure until Mercator's independent development." *Flattening*, 48.

101 *hundreds of miles from his intended destination*: Cf. Anson's famous circumnavigation, when his "ships almost ran ashore on the rocks of the Chilean coast at a moment when the sailing-masters estimated that they were nearly three hundred miles out to sea." Glyndwr Williams, *The Great South Sea: English Voyages and Encounters, 1570–1750* (New Haven, CT: Yale University Press, 1997), 10.

102 *Columbus used Ptolemy's calculations*: Boorstin, *Discoverers*, 99.

104 *bleed across the whole world*: Monmonier, *Lie with Maps*, 96.

104 *"south until the butter melts"*: A sixteenth-century saying about sailing to the West Indies from Europe, quoted in William Galvani, *Mainsail to the Wind: A Book of Sailing Quotations* (Dobbs Ferry, NY: Sheridan House, 1999), 71.

105 *inevitable land mass distortions*: Crane, *Mercator*, 289–90.

105 *distortion is coming for you*: Ibid., 290.

105 *imaginary Christian nation . . . hiding out in Ethiopia*: It was moved to Ethiopia (at this time, Abyssinia) in Mercator's era. After evidence against its existence came to light, the last gasp of the myth relied on scholars asserting that the name "Prester John" was a corruption of certain Ethiopian royal titles. Almeida is the first to dismantle that argument, saying that "there is no foundation whatever" for it. Quoted in Robert Silverberg, *The Realm of Prester John* (Athens, OH: Ohio University Press, 1972), 319–21. Also see Keagan Brewer, comp. and trans., *Prester John: The Legend and Its Sources* (Surrey: Ashgate, 2015).

106 *instead of for our yahoos*: Questions on Yahoo Answers from May 4, 2012 (https://answers.yahoo.com/question/index?qid=20120504132447AAweXOA); January 26, 2008 (https://answers.yahoo.com/question/index?qid=20080126235022AAPDIDx); and October 30, 2012 (https://au.answers.yahoo.com/question/index?qid=20121030224201AAnKzZZ). The top-rated answer to the last question is "Can you be more Pacific?" People from 2517, go ahead and judge us for that answer.

Chapter 5: Bad Shakespeare

107 *"To die, to sleep"*: William Shakespeare, *Hamlet*, act 3, scene 1, lines 59–63. Quoted in *The Norton Shakespeare*, ed. by Stephen Greenblatt (New York: W.W. Norton and Company, 1997), 1705.

107 *"is that all?"*: William Shakespeare, *Hamlet* Q1, transcription from the British Library copy (except the spelling out of "I" to "Aye," as in some later editions, to avoid confusion), D4^{v}. See http://www.quartos.org/lib/XMLDoc/viewXML.php?path=ham-1603–22275x-bli-c01.xml.

107 *the "bad" quarto*: The term was first used by A. W. Pollard in *Shakespeare's Folios and Quartos* (London: Methuen, 1909).

108 "found by me in a closet": The imprint was bound in a collection of Shakespeare plays, one reason it may have taken so long to discover it. Sir Henry Bunbury, ed., *The Correspondence of Sir Thomas Hanmer, Bart, Speaker of the House of Commons, with a Memoir of his Life* (London: Edward Moxon, 1838), 80.

108 "at a tidy profit": Zachary Lesser, *Hamlet After Q1: An Uncanny History of the Shakespearean Text* (Philadelphia: University of Pennsylvania Press, 2015), 3. This friend of Charles Dickens was the well-known book collector William Cavendish, Sixth Duke of Devonshire.

109 presumably repentant British Museum: Ibid., 16.

109 best-selling playwright: Lukas Erne, *Shakespeare and the Book Trade* (Cambridge, UK: Cambridge University Press, 2013), 21.

109 even if they weren't his: Ibid., 68.

110 "music be the food of love": William Shakespeare, *Twelfth Night*, act 1, scene 1, line 1. Quoted in *Norton Shakespeare*, 1768.

110 "I have done thy mother": William Shakespeare, *Titus Andronicus*, act 4, scene 2, lines 75–76. Quoted in *Norton Shakespeare*, 414.

110 "if an editor likes a reading": R. B. McKerrow, *The Treatment of Shakespeare's Text by His Earlier Editors, 1709–1786* (Freeport, NY: Books for Libraries Press, 1970), 28.

110 "authentic" Shakespeare play: Spoilers: *authentic* is in quotes because it's not a terribly practical way to understand Shakespearean authorship, as demonstrated by the historical context described in this chapter.

110 "historically based editorial practice": Kastan, *Shakespeare and the Book*, 100.

111 start tossing them out: Even the great libraries weren't immune, such as the Bodleian "discarding" its First Folio after obtaining a Third Folio. See Erne, *Shakespeare and the Book Trade*, 192.

111 seemed to show metrical errors: In fact these verses were, as later scholars have pointed out, "entirely acceptable when accorded their more flexible Elizabethan and Jacobean pronunciation." Margreta de Grazia, *Shakespeare Verbatim: The Reproduction of Authenticity and the 1790 Apparatus* (Oxford: Clarendon Press, 1991), 62–63.

111 fifteen hundred "degraded": Ibid., 197.

112 "Shakespearean controversy has given birth": Thomas R. Lounsbury, *The First Editors of Shakespeare (Pope and Theobald): The Story of the First Shakespearean Controversy and the Earliest Attempt at Establishing a Critical Text of Shakespeare* (London: David Nutt, 1906), xi.

112 *his publisher turned traitor*: Andrew Murphy, "The Birth of the Editor," in *A Concise Companion to Shakespeare and the Text*, ed. Andrew Murphy (Hoboken, NJ: Wiley-Blackwell, 2007), 96. It should be noted that Theobald *also* went on to edit (and alter) Shakespeare according to his own morally inflected judgments.

112 *"suspicions of depravity"*: As quoted in De Grazia, who emphasizes Theobald's suspiciously moralizing vocabulary: Depravities, Deformities, Blemishes. *Verbatim*, 198–99.

112 *most blasé . . . is the man himself*: Erne challenges this interpretation using circumstantial evidence, but there is no direct evidence.

112 *Plays were usually sold to a theater team*: David Grote, *The Best Actors in the World: Shakespeare and His Acting Company* (Westport, CT: Greenwood Press, 2002), 22.

112 *free to make changes to the text*: Tiffany Stern, *Making Shakespeare: From Stage to Page* (London: Routledge, 2004), 57.

113 *"nothing too seditious or blasphemous"*: Ibid., 154.

113 *"c's, her u's, and ['n'] her t's"*: William Shakespeare, *Twelfth Night*, act 2, scene 5, lines 77–78. Quoted in *Norton Shakespeare*, 1791.

113 *euphemism for vagina*: Gordon Williams, *Shakespeare's Sexual Language: A Glossary* (London: Continuum, 2006), 219.

113 *abridged traveling version*: This theory, relying on a sort of "country bumpkin" argument, has seemed less convincing in recent years. Leah S. Marcus, *Unediting the Renaissance: Shakespeare, Marlowe, Milton* (London: Routledge, 2002), 150–52.

113 *memorial reconstruction*: While not first proposed by W. W. Greg, this theory was popularized by him in his introduction to his 1910 *Shakespeare's Merry Wives of Windsor*. In 1915, Henry David Gray argued that the bit actor playing Marcellus was responsible for the memorial reconstruction of the "bad" *Hamlet*, in "The First Quarto *Hamlet*," *Modern Language Review*, 10 (1915): 171–80.

113 *"corrupt and mangled (copied only by ear)"*: Thomas Heywood, "To the Reader," in *The Rape of Lucrece* (London: Printed [by E. Allde] for I. B[usby] and are to be sold [by Nathaniel Butter], 1608), sig. A2^{r}.

113 *questionable just how widespread*: In correspondence with the authors, Adam Hooks noted this practice was more common for sermons, and that in fact it is one more way for editors to blame other forces for the less pleasing parts of Shakespeare's plays.

114 *"make so bold with his name"*: Thomas Heywood, *Apology for Actors* (London: Nicholas Okes, 1612), sig. G4^{r}.

114 "deemed worthy": Erne, *Shakespeare and the Book Trade*, 73.

114 "a name to make money": Ibid., 59.

114 removed Shakespeare's name: Adam Hooks pointed out in correspondence that Colin Burrow makes the opposite argument in his Oxford edition of Shakespeare. Welcome to Shakespearean criticism.

115 "how little we know": James J. Marino, *Owning William Shakespeare: The King's Men and Their Intellectual Property* (Philadelphia, University of Pennsylvania Press, 2011), 72.

115 eighty-three variants: Margreta de Grazia and Peter Stallybrass, "The Materiality of the Shakespearean Text," in *Shakespeare and the Editorial Tradition*, ed. Stephen Orgel and Sean Keilen (London: Taylor and Francis, 1999), 19.

115 condemned to have his name misspelled: Randall McLeod, "Un 'Editing' Shak-speare," *SubStance* 10/11 (vol. 10, no. 4; vol. 11, no. 1), Issue 33–34: *Books: On and About* (1981/1982): 4546. Note that this practice doesn't specifically apply to Roman letters; however, the principle remains that our spelling of Shakespeare derives from printed matter, not manuscript—including Shakespeare's own writing.

115 "authors do not write books": Roger Stoddard, "Morphology and the Book from an American Perspective," *Printing History* 17, vol. 9, no. 1 (1987): 4.

116 "fills bibliographers with horror": Thomas L. Berger, "Shakespeare Writ Small" in *Concise Companion to Shakespeare and the Text*, ed. Andrew Murphy (Hoboken, NJ: Wiley-Blackwell, 2007), 68.

116 follically challenged wild animals: Stern, *Making Shakespeare*, 151. This is also where we note that the editorial choice of standardizing spelling and punctuation for ease of reading—as we do in this book, have mercy on our souls—can actually erase important textual evidence. The white beares/white beards mistake can't be seen if you normalize the spelling to "bears."

116 had just run out of periods: Ibid., 152.

116 gets a name change to "Bastard": Ibid., 152–53.

117 "collection" of great Elizabethan dramas: Jeffrey Todd Knight makes the distinction that most "collections" of bound plays from this era were determined by a reader buying up each play separately and then binding them together at a point of his choosing. However, some publishers learned early on that there could be benefits to printing certain plays at the same time, when they could sell the two or three either individually or together as a sort of collection. See Tara L. Lyons, "Serials, Spinoffs, and Histories: Making 'Shakespeare' in Collection before the Folio," *Philological Quarterly* 91, no. 2 (2012):

185–220; and Jeffrey Todd Knight, "Making Shakespeare's Books: Assembly and Intertextuality in the Archives," *Shakespeare Quarterly* 60 (2009): 304–40. All that said, the 1619 collection was the largest group up to that point of Shakespeare plays printed more or less consecutively and under the supervision of a single publisher (Jaggard).

117 *ale, beer, and "other gross wares"*: We know this because, of course, someone complained about it. See the August 5, 1554, Proclamation of the Lord Mayor, quoted in William Benham, *Old St. Paul's Cathedral* (London: Seeley and Co., 1902), 47.

118 *respectable heart of the London book trade*: See Peter W. M. Blayney, *The Bookshops in Paul's Cross Churchyard* (London: Bibliographical Society, 1990) for a full treatment.

118 *went blind from his STD or its "cure"*: Jillian Linster, "When 'Nothing' Goes Missing: The Impotent Censorship of Helkiah Crooke's *Mikrokosmographia*," paper presented on March 10, 2013, https://crookebook.files.wordpress.com/2013/03/crooke-images-paper-3–10–13.pdf.

118 *pornographic indecency*: Ibid.

118 *"vaginal cleft in full anatomical detail"*: Ibid.

119 *first comprehensive anatomy volume*: Ibid.

119 *remain free from moral corruption*: Like when Samuel Pepys purchased a "mighty lewd book," which "doth me no wrong to read for information sake." Right. He adds, "(but it did hazer my prick para stand all the while and una vez to decharger)," mixing in foreign languages for the scandalous bits, in the grand tradition that Edward Gibbon would follow one hundred fifty years later. Gibbon made copious use of Latin in his *History of the Decline and Fall of the Roman Empire* (1776–89) in these cases: "My English text is chaste, and all licentious passages are left in the decent obscurity of a learned language." Samuel Pepys, diary entry for February 9, 1668, quoted in *The Illustrated Pepys: Extracts from the Diary*, ed. by Robert Latham (Berkeley: University of California Press, 1978), 173–74. Gibbon, *Vindication* (London: 1779).

119 *"burn it wherever he found it"*: Linster, "Nothing."

119 *The college yelled at her for a while*: Ibid.

120 *yearly salary of around two hundred pounds*: Andrew Gurr, *The Shakespeare Company, 1594–1642* (Cambridge, UK: Cambridge University Press, 2004), 115.

120 *cost of attending university*: J. Payne Collier, *Memoirs of the Principal Actors in the Plays of Shakespeare* (London: Printed for the Shakespeare Society, 1846), 147.

120 *"great living wealth and power"*: Quoted in E. K. Chambers, *The Elizabethan Stage*, vol. 2 (Oxford: Clarendon Press, 1923), 323.

121 *king's personal eye candy*: Gurr, *Shakespeare Company*, 51.

121 *illegal for anyone to publish a play of Shakespeare*: Transcribed in William A. Jackson, ed., *Records of the Court of the Stationers' Company: 1602–1640* (London: Bibliographical Society, 1957), 110.

122 *branded the Pavier Quartos*: Technically there are only five imprints of "Printed for T.P.," but *The Whole Contention* was a combination of two plays, therefore making the Pavier attribution appropriate for six in total.

122 *rejected from the canon as apocryphal*: See Peter Kirwan, *Shakespeare and the Idea of Apocrypha* (Cambridge, UK: Cambridge University Press, 2015).

122 *smashed the surplus theory . . . watermarks . . . "devices"*: W. W. Greg, "On Certain False Dates in Shakespearean Quartos," *The Library*, New Series, 9, no. 34 (1908): 122–23 and 131.

123 *Jaggard Quartos*: Zachary Lesser and Peter Stallybrass, "Shakespeare Between Pamphlet and Book," in *Shakespeare and Textual Studies*, ed. Margaret Jane Kidnie and Sonia Massai (Cambridge, UK: Cambridge University Press, 2015), 133.

124 *branded literary pirates*: A. W. Pollard, *Shakespeare's Fight with the Pirates and the Problems of the Transmission of His Text* (London: A. Moring, 1917).

124 *McKerrow device #283*: Ronald B. McKerrow, *Printers' and Publishers' Devices in England and Scotland, 1485–1640* (London: Bibliographical Society, 1913). Yes, there are literally entire dictionaries compiled of printers' devices.

124 *known who printed the item*: Marino, *Owning Shakespeare*, 108–9.

125 *ornately designed "Go hang yourselves"*: Particularly argued by Marino in ibid., 114.

125 *oversaw all printing in London*: See Peter W. M. Blayney, *The Stationers' Company and the Printers of London, 1501–1557* (Cambridge, UK: Cambridge University Press, 2013).

125 *patents . . . biased royal meddling*: Joseph Loewenstein, *The Author's Due: Printing and the Prehistory of Copyright* (Chicago, IL: University of Chicago Press, 2002), 29. See, for example, the fabulous career of John Wolfe, whose business was raided again and again for piracies. He kept fighting the existing printing laws, even from jail, and in the end landed a cushy gig as the company beadle of the stationers: the man in charge of the search, seizure, and arrest of anyone breaking the laws of the book trade.

126 *"printed without the consent"*: Jackson, *Records*, 110.

126 *inherited or traded or sold*: See Loewenstein, *Author's Due*, esp. p. 120, about business deals with monopolies.

126 *Heywood's* Woman Killed: See Lesser and Stallybrass, "Pamphlet," 127.

The inclusion of a Heywood play plus two apocryphal Shakespeare works (*Sir John Oldcastle* and *Yorkshire Tragedy*) and the high degree of variability in the order of the works in surviving bound copies suggest that "Jaggard appears to be going to great lengths to convince someone (who?) that what might appear to be a 'Shakespeare collection' is actually highly variable—and therefore unlikely to be the work of a bookseller [but of private owners]" (ibid., 132). Note that Lesser and Stallybrass aren't willing to commit to the idea that Jaggard was specifically trying to fool the King's Men players, although they are not explictly discounted.

126 rights . . . obtained for publication: Marino, *Owning Shakespeare*, 130–31.

126 "clotpoles": Cf. William Shakespeare, *Troilus and Cressida*, act 2, scene 1, line 118-ish.

127 his own dear "orphans": John Heminges and Henry Condell, dedication to *Mr. William Shakespeares comedies, histories, and tragedies. Published according to the true originall copies* (London: Isaac Jaggard, and Ed. Blount, 1623), A2[v].

127 "Do so, but buy it first": Heminges and Condell, "To the Great Variety of Readers," in ibid., A3[r].

127 making money off the process: In fairness to their mercenary urges, it's more likely Heminges and Condell would have made their money indirectly, from encouraging Shakespeare's reputation as a playwright, rather than from actual sales of the book.

127 dead husband's shareholdings: Gurr, *Shakespeare Company*, 230. Another classic example of learning biographical details through the record of lawsuits.

127 owned a country home: Grote, *Best Actors*, 218.

127 February 1622, printing began: Peter W. M. Blayney, *The First Folio of Shakespeare* (Washington, DC: Folger Library, 1991), 5.

127 "incomparably the most important work in the English language": E. V. Unger and W. A. Jackson, eds., *The Carl H. Pforzheimer Library: English Literature, 1475–1700*, vol. 3 (New York: Morrill Press, 1940), 935.

128 exist only because of this collection: Blayney, *First Folio*, 1.

128 much larger, taller folio: Amanda Mabillard, *Shakespeare in Print*, Shakespeare Online, http://www.shakespeare-online.com/biography/shakespeare inprint.html.

128 you can still see the delicate stab-holes: Roger Chartier and Peter Stallybrass, "What Is a Book?" in *Cambridge Companion to Textual Scholarship*, ed. Neil Fraistat and Julia Flanders (Cambridge, UK: Cambridge University Press, 2013), 196.

129 *"frauds and stealths of injurious imposters"*: Heminges and Condell, "To the Great Variety of Readers," A3ʳ.

129 *running the everyday operations*: Blayney, *First Folio*, 2, 4.

129 *"literary arbiter of taste"*: Sonia Massai, "Edward Blount, the Herberts, and the First Folio," in *Shakespeare's Stationers*, ed. Marta Straznicky (Philadelphia: University of Pennsylvania Press, 2013), 138.

130 *"no certainty of recovering their considerable investment"*: Kastan, *Shakespeare and the Book*, 78.

130 *admitted into the folio club*: Ibid., 51.

130 *"but for all time"*: Ben Jonson, "To the Memory of My Beloved, the Author Mr. William Shakespeare: And What He Hath Left Us," in *Mr. William Shakespeares Comedies, Histories, and Tragedies*, A4ᵛ.

131 *attempted to raise dramatic arts*: See Stephen Orgel, "Shakespeare and the Kinds of Drama," *Critical Inquiry* 6, no. 1 (1979): 107–23; and Andrew Murphy, *Shakespeare in Print: A History and Chronology of Shakespeare Publishing* (Cambridge, UK: Cambridge University Press, 2003), 35.

131 *"What others call a play you call a work"*: Anonymous, quoted in *Wits Recreations* (London: R.H. for Humphrey Blunden, 1640), G3ᵛ. The response of a friend of Jonson's is also recorded: "Ben's plays are works, when others' works are plays." That last guy definitely lost the rap battle.

131 *five shillings off the sticker price*: Blayney, *First Folio*, 28–32.

131 *96,000 loaves of bread*: Murphy, *Shakespeare in Print*, 51 and 422, referencing approximations suggested by Anthony West in "Sales and Prices of Shakespeare First Folios: A History, 1623 to the Present," Part One, *Papers of the Bibliographical Society of America* 92, no. 4 (1998): 465–528.

131 *first recorded sale*: Blayney, *First Folio*, 25.

132 *"a respectable performance"*: Ibid., 32.

132 *bring upon themselves a swashbuckler's fate*: "We're avoiding the curse," said the man in charge of the renovations. "Bard's 'Cursed' Tomb Is Revamped," *BBC News*, May 28, 2008, http://news.bbc.co.uk/2/hi/uk_news/england/coventry_warwickshire/7422986.stm.

132 *skull was likely stolen from its grave*: Sarah Kaplan, "Shakespeare's Skull Probably Isn't in His Grave," *Washington Post*, March 25, 2016, https://www.washingtonpost.com/news/morning-mix/wp/2016/03/25/shakespeares-skull-probably-isnt-in-his-grave/.

133 *"set the table on a rore"*: William Shakespeare, *Hamlet*, act 5, scene 1, lines 171 and 175–77. Quoted in *Norton Shakespeare*, 1744.

133 "the creator of Shakespeare": Kastan, *Shakespeare and the Book*, 78.

134 "a blot in his papers": Heminges and Condell, "To the great Variety of Readers," in *Mr. William Shakespeares Comedies, Histories, & Tragedies*, A3r.

134 "pusles the brain . . . others that we know not of": Q1, 1603, "bad" *Hamlet*, E1r.

Chapter 6: Benjamin Franklin Makes It Rain

135 "new suit" . . ."five pounds sterling": Benjamin Franklin, *The Autobiography of Benjamin Franklin*, ed. Leonard W. Labaree, Ralph L. Ketcham, Helen C. Boatfield, and Helene H. Fineman, with foreword by Edmund S. Morgan (New Haven, CT: Yale University Press, 2003), 81.

135 that *kind of a wife*: Hugh Amory, "Reinventing the Colonial Book," in *A History of the Book in America*, vol. 1: *The Colonial Book in the Atlantic World*, ed. Hugh Amory and David D. Hall (Chapel Hill: University of North Carolina Press, 2009), 51.

136 "no such sum" . . ."mortgage" . . ."not approve": Franklin, *Autobiography*, 127.

136 "printing business was not a profitable one" . . ."soon follow them": Ibid.

136 "poor one" . . ."not to expect money with a wife": Ibid., 128.

137 "intrigues with low women": Ibid.

137 "good and faithful": Ibid., 129.

137 follow the money: We are indebted to Michael F. Suarez, S.J., and his course at the University of Virginia's Rare Book School, *Teaching the History of the Book*, for drilling this concept into a permanent feature of Rebecca's brain.

138 no gambling, no drinking, and absolutely no fornicating: Ralph Frasca, *Benjamin Franklin's Printing Network: Disseminating Virtue in Early America* (Columbia: University of Missouri Press, 2006), 22–23.

138 "'obliged to take the lowest people'": Ibid., 32.

139 the "indentured" part of indentured servitude: James N. Green and Peter Stallybrass, *Benjamin Franklin: Writer and Printer* (New Castle, DE: Oak Knoll Press, 2006), 50.

139 "talkative when drunk and remarkably stupid": May 2, 1765. Quoted in Frasca, *Franklin*, 27.

139 "which I took extremely amiss": Franklin, *Autobiography*, 68.

139 "the exquisite pleasure": Ibid.

139 "a state I never much admir'd": Silence Dogood, essay 2, "Sir, Histories of Lives Are Seldom Entertaining . . ." in *The New-England Courant* 37, April

9–16, 1722. Accessed online via the Massachusetts Historical Society Collections Online Resources, http://www.masshist.org/online/silence_dogood/doc-viewer.php?item_id=634&pid=6.

140 *"learning has brought disobedience and heresy"*: The governor, a true politician, spoke these words while answering a question in a way completely unrelated to the actual question: "What course is taken about instructing the people, within your government in the Christian religion, and what provision is there made for the paying of your ministry?" Quoted in William Waller Hening, *The Statutes at Large; Being a Collection of all the Laws of Virginia, from the First Session of the Legislature in the year 1619*, vol. 2 (Richmond: Samuel Pleasants, Junior, 1810), 517.

141 *90 percent of all printing in the colonies*: David D. Hall, "The Atlantic Economy in the Eighteenth Century," in *A History of the Book in America*, vol. 1: *The Colonial Book in the Atlantic World*, ed. Hugh Amory and David D. Hall (Chapel Hill: University of North Carolina Press, 2009), 155.

141 *you really couldn't blame him*: Amory, "Reinventing," 26.

142 *"you shall print nothing without allowance"*: Quoted in James N. Green, "The Book Trade in the Middle Colonies, 1680–1720," in *A History of the Book in America*, vol. 1: *The Colonial Book in the Atlantic World*, ed. Hugh Amory and David D. Hall (Chapel Hill: University of North Carolina Press, 2009), 204.

142 *economic motivation for a free press*: See Franklin's "Apology for Printers," *The Pennsylvania Gazette*, May 27, 1731, especially points 6 and 7. An excerpt: ". . . regarding it only as the matter of their daily labour: They print things full of spleen and animosity, with the utmost calmness and indifference . . . since in the way of their business they print such great variety of things opposite and contradictory."

142 *a press that was open to everyone*: James N. Green, "English Books and Printing in the Age of Franklin," in *A History of the Book in America*, vol. 1: *The Colonial Book in the Atlantic World*, ed. Hugh Amory and David D. Hall (Chapel Hill: University of North Carolina Press, 2009), 256.

142 *"a very flimsy scheme"*: Franklin, *Autobiography*, 70.

143 *"got a naughty girl with child"*: Ibid., 71.

143 *press was "shatter'd,"*: Ibid., 78.

143 *more than one printer in a Colonial city*: Green, "English Books," 248.

144 *"better dressed than ever"*: Franklin, *Autobiography*, 81.

144 *"a piece of eight to drink"*: Ibid., 81–82.

144 *"could never forget or forgive"*: Ibid., 82.

144 *"I will do it my self"*: Ibid., 86.

145 *"see that everything was good"*: Ibid., 87.

145 *"no credit to give"*: Ibid., 94.

145 *"drank every day a pint"*: Ibid., 100.

146 *"alehouse boy"*: Ibid.

146 *"the Water-American"*: Ibid., 99–100.

146 *contemporary cartoons referred to them as "horses"*: There's a very famous illustration in particular that appeared in the *Grub-Street Journal* on October 20, 1732, which shows pullers as horses and compositors as asses.

146 *compositors were later called "monkeys"*: Cf. Honoré de Balzac, *Lost Illusions. Part First. The Two Poets. I. A Printing-House in the Provinces, Where Monkeys Are Compositors and Pressmen Are Bears*; and William Savage, *Dictionary of the Art of Printing*: "The compositers [*sic*] are jocosely called Galley Slaves: Because allusively they are as it were bound to their Gallies. And the Press-men are jocosely called Horses: Because of the hard labour they go through all day long" (London: Longman, Brown, Green, and Longmans, 1841), 12.

146 *"more strength than a quart of beer"*: Franklin, *Autobiography*, 100.

147 *"uncommon quickness"*: Ibid., 101.

147 *compositors were paid by the number of lines*: Of course this varied by time and location, but it is generally supported by documentary evidence in England during this period. For further discussion see Keith Maslen, "Jobbing Printing and the Bibliographer: New Evidence from the Bowyer Ledgers" in *An Early London Printing House at Work: Studies in the Bowyer Ledgers* (New York: Biographical Scoeity of America, 1993), 139–52.

147 *"poor devils keep themselves always under"*: Franklin, *Autobiography*, 100.

148 *"hot water-gruel"*: Ibid., 101.

148 *his journeyman salary*: Green and Stallybrass, *Franklin*, 29–30.

148 *weren't even close to self-sufficient*: David D. Hall, "Introduction, Part One: Some Contexts and Questions," in *A History of the Book in America*, vol. 1: *The Colonial Book in the Atlantic World*, ed. Hugh Amory and David D. Hall (Chapel Hill: University of North Carolina Press, 2009), 8.

148 *American Bible was suspect*: Green and Stallybrass, *Franklin*, 71.

149 *complete Bible in English would proudly be published*: It's worth noting that, beyond the economic problems, there was also a legal obstacle to an American Bible: the royal patent was held by a printer in London. According to the printer Isaiah Thomas, there was an earlier Bible printed in America, around 1752. However, it bore a false London imprint. To date, no copies of it have been discovered.

149 *just replacement quotation marks*: William C. Miller, *Benjamin Franklin's Philadelphia Printing 1728–1766: A Descriptive Bibliography* (Philadelphia, PA: American Philosophical Society, 1974), xxxiii.

149 *first type foundry established on American soil*: John Bidwell, "Printers' Supplies and Capitalization," in *A History of the Book in America*, vol. 1: *The Colonial Book in the Atlantic World*, ed. Hugh Amory and David D. Hall (Chapel Hill: University of North Carolina Press, 2009), 169.

149 *"take salt water out of books"*: Quoted in Green, "English Books," 262.

149 *commonly bound in sheepskin*: Miller, *Philadelphia Printing*, xlix.

150 *issued "stitched"*: Ibid., xlvii.

150 *"AFTER PROFIT"*: Quoted in Johan Gerritsen, "Printing at Froben's: An Eye-Witness Account," *Studies in Bibliography* 44 (1991): 150. Cf. James Raven, *Business of Books* (New Haven, CT: Yale University Press, 2007), 6: "The financing and business arrangements of the early modern book trade were surprisingly sophisticated, and very often profit, not ideology, proved the more compelling."

150 *size of a briefcase*: Green and Stallybrass, *Franklin*, 64.

151 *"character and credit"*: Franklin, *Autobiography*, 119.

151 *"ostentatious, almost superhuman speed"*: Green and Stallybrass, *Franklin*, 32.

151 *"no way entertaining"*: Franklin, *Autobiography*, 119. *Stuff* is almost an appropriate word here—it's the technical term for the pulpy liquid that papermakers pulled from their vats to create paper.

151 *"enough for America"*: Ibid., 67.

152 *"what relates to A"*: *Pennsylvania Gazette*, December 24, 1728. Today, only a few copies of this newspaper are known to exist, all in institutions.

152 *"will probably be fifty years"*: Franklin's own *Pennsylvania Gazette*, October 2, 1729.

153 *text he printed was a mess*: Franklin, *Autobiography*, 121, says it was printed "in a coarse and blundering manner." See also Green and Stallybrass, *Franklin*, 37.

153 *"now and then . . . satirize a little"*: The quote in which we reveal our own debt to Franklin. "The Busy-Body, No. 4," *American Weekly Mercury*, February 25, 1729.

154 *"the life of a paper."*: This was James Parker, in a 1769 letter to Franklin. Quoted in Frasca, *Printing Network*, 142.

154 *"better distributer of advertisements"*: Franklin, Autobiography, 126–27.

154 *"bribing the riders"*: Ibid., 127.

154 inexactitude: Ibid., 172.

154 subscriptions had jumped: Green and Stallybrass, 36–37.

155 more than 50 percent of Franklin's income: This is during the period of his partnership with Hall, cf. Miller, *Philadelphia Printing*, xxx.

155 paper from England: Ibid., xxxviii.

155 "forcing Franklin to import paper": Green and Stallybrass, *Franklin*, 29, 40.

155 a new local mill: Ibid., 40.

155 eighty-three tons of rags: Ibid., 40.

156 "more copies of almanacs were sold than all other": James Raven, *The Publishing Business in Eighteenth-Century England* (Suffolk: Boydell Brewer, 2014), 201–2.

156 "two of them are dead": *Poor Richard's Almanack*, June 1738; December 1732 (the first of the *Poor Richard Almanacs*); and July 1735.

156 "copper-plate press": Franklin, *Autobiography*, 112.

156 "no one else in America": Green and Stallybrass, *Franklin*, 29.

156 dominating the transatlantic book trade: Ibid., 45.

157 "wrapping papers for soap and tobacco": Ibid., 47–49,

157 four thousand salad oil advertisements: Raven, *Publishing*, 116.

157 whatever issue is fashionable at the moment: Andrew Pettegree, *Reformation and the Culture of Persuasion* (Cambridge, UK: Cambridge University Press, 2005), 152.

157 receipts, lottery tickets, advertisements: Green and Stallybrass, *Franklin*, 49.

158 "most delightful carpet advertisement ever created": Robert N. Essick, *William Blake, Printmaker* (Princeton, NJ: Princeton University Press, 1980), 81.

158 printer's bread and butter: Pettegree, *Reformation*, 153.

158 substantial source of income: See Green, "English Books", 266; and Scott E. Casper and Joan Shelley Rubin, "The History of the Book in America," in *The Book: A Global History*, ed. Michael F. Suarez, S.J., and H. R. Woudhuysen (Oxford: Oxford University Press, 2013), 685.

158 "regular, quick turnover": Raven, *Publishing*, 45.

158 literally stopped the presses: Green and Stallybrass, *Franklin*, 47.

158 chocolate, whalebone, pickled sturgeon: Amory, "Reinventing," 45.

158 "wealthiest in all the colonies": Green and Stallybrass, *Franklin*, 45.

159 "after getting the first hundred pound": Franklin, *Autobiography*, 180–81.

159 "sent one of my journeymen to Charleston": Ibid., 166.

159 "with Franklin at the center": This quote is from Green and Stallybrass, *Franklin*, 42; but for a book-length treatment, see Frasca, *Printing Network.*

160 gathering place for that community: Green, "English Books," 271.

160 "vehicles of discussion": Isaiah Thomas, *The History of Printing in America*, vol. 2 (Worchester, MA: Isaiah Thomas Jr., 1810), 403, quoting Miller, *Retrospect of the Eighteenth Century.*

160 carried weight in South Carolina: Frasca, *Printing Network*, 72–73.

160 "could happen to them": Edmund S. Morgan, *Benjamin Franklin* (New Haven, CT: Yale University Press, 2003), 207.

160 "symbol of colonial opposition": Frasca, *Printing Network*, 149.

161 no one revolted: Raven, *Publishing*, 69.

161 pre-Revolutionary furor: Hall, "Atlantic Economy," 156. See also Charles E. Clark, "Early American Journalism: News and Opinion in the Popular Press," in *A History of the Book in America*, vol. 1: *The Colonial Book in the Atlantic World*, ed. Hugh Amory and David D. Hall (Chapel Hill: University of North Carolina Press, 2009), 361.

161 melted it down to form bullets: Thomas, *History*, 313.

161 agent for George Washington's Culper Spy Ring: Unsurprisingly, the documentation for Revolutionary-era spy efforts is sparse, but most scholars today agree that Rivington was an agent, if they can't quite agree on the dates he worked for General Washington. See Todd Andrlik, "James Rivington: King's Printer and Patriot Spy?" *Journal of the American Revolution*, March 3, 2014, https://allthingsliberty.com/2014/03/james-rivington-kings-printer-patriot-spy/; and Kara Pierce, "A Revolutionary Masquerade: The Chronicle of James Rivington," *Binghamton Journal of History*, last updated August 24, 2010, https://www.binghamton.edu/history/resources/journal-of-history/chronicles-of-james-rivington.html.

161 "starting the American Revolution": Morgan, *Franklin*, 15.

162 "expense of heating the iron": As recounted in Frasca, *Printing Network*, 152.

162 "mere mechanics": Stephen Botein, "'Meer Mechanics' and an Open Press: The Business and Political Strategies of Colonial American Printers," *Perspectives in American History* 9 (1975), 222.

Chapter 7: Angelic Visions and Deadly Terrors

163 "photographer of his day": Febvre and Martin, *Coming of the Book*, 102.

163 reproduce cat hair: William M. Ivins Jr., *Prints and Visual Communication* (Cambridge, MA: MIT Press, 1969), 70.

164 *"be enslaved by another man's"*: William Blake, *Jerusalem the Emanation of the Giant Albion* (London: W. Blake, 1804), 10.

164 *wax, needles, and nitric acid*: Or, as was often the case, a combination of all three. See William M. Ivins Jr., *How Prints Look*, revised by Marjorie B. Cohn (Boston: Beacon Press, 1987), 46. See also Anthony Griffiths, *Prints and Printmaking: An Introduction to the History and Techniques* (Berkeley: University of California Press, 1996), for a survey of illustration processes, including relief, intaglio, planographic, and photomechanical processes.

165 *"cheap, crude, and effective"*: R. W. Scribner, *For the Sake of Simple Folk: Popular Propaganda for the German Reformation* (Cambridge: Cambridge University Press, 1981), 5.

165 *"Even Milton and Shakespeare"*: William Blake, *Prospectus*, October 10, 1793. Because it's an individually engraved piece of ephemera, and therefore subject to typical devastation, no copies survive. We have the text thanks to Alexander Gilchrist, *Life of William Blake*, 2 vols. (London: Macmillan, 1863), 2:263.

165 *turpentine, asphaltum [black petroleum], and linseed oil*: Joseph Viscomi, *Blake and the Idea of the Book* (Princeton, NJ: Princeton University Press, 1993), 56.

165 *dead at the time*: This story is told in Gilchrist, *Life*, 1:69.

166 *translate into significant changes*: See Jerome J. McGann, *The Textual Condition* (Princeton, NJ: Princeton University Press), 199. Elaborating on Thomas G. Tanselle's work, McGann argues that "every documentary or bibliographical aspect of a literary work is meaningful, and potentially significant," including its material aspect, and that therefore "all editing is an act of interpretation" (78, 27).

166 *breathtaking in its original state*: Because we can't all march into the Huntington and browse through original illuminated printing by Blake, we offer you the gift of the Blake Archive, one of the best websites dedicated to making rare books on a specific topic digitally accessible: blakearchive.org. Also note that the hand-coloring on Blake's plates did not remain uniform across copies, so while we describe blues, reds, greens, and yellows here, there are copies that are mostly yellow and blue, or yellow and green. At the Blake Archive you can compare these various copies.

168 *"exploits this convention"*: Robert N. Essick, *William Blake at the Huntington* (New York: Henry N. Abrams, 1994), 52.

168 *"promised sloth, debauchery"*: Altick, *Common Reader*, 33.

169 *"bereav'd of light"*: Poem transcribed from the Yale Center for British Art's copy of William Blake, *Songs of Innocence*, objects 29 and 30. Available online at the Blake Archive.

169 *"black 'guardian angel'"*: W. J. T. Mitchell, *Blake's Composite Art: A Study of the Illuminated Poetry* (Princeton, NJ: Princeton University Press, 1978), 12–13.

175 *"neither can nor ought to exist"*: *The British Critic*, September 1796, quoted in G. E. Bentley Jr., *The Stranger from Paradise* (New Haven, CT: Yale University Press, 2001), 117.

175 *deerstalker . . . an image by Sidney Paget*: Footnote 3 by Klinger in Arthur Conan Doyle, "Silver Blaze," in *The New Annotated Sherlock Holmes*, vol. 1, ed. with a foreword and notes by Leslie S. Klinger (New York: W.W. Norton and Company, 2005), 388.

176 *"chiefly by his close friends"*: Bentley, *Stranger*, 117.

176 *"set you a-screaming"*: Thomas Sadler, ed., *Diary, Reminiscences, and Correspondence of Henry Crabb Robinson: Second Edition*, 3 vols. (London: Macmillan, 1869), 2:383.

176 *granting glorious visions*: For more jaw-dropping anecdotes on visions, see G. E. Bentley Jr., ed., *William Blake: The Critical Heritage* (London: Routledge, 2002), 36–39.

177 *"inspired by a god"*: A breakdown of the Greek goes *en* ("in") + *theos* ("god") + *ousia* ("essence").

177 *"ordinary unemphatic tone"*: Sadler, *Reminiscences*, 2:302.

177 *"an unfortunate lunatic"*: Anonymous review of Blake's *Descriptive Catalogue* in the September 17, 1809, issue of *The Examiner*, written by Robert Hunt. Quoted in Bentley, *Stranger*, 332–33.

177 *"his abstract habits"*: John Flaxman Jr., letter of December 1, 1805. Quoted in Bentley, *Stranger*, 380.

177 *"always in Paradise"*: Seymour Kirkup, letter of March 25, 1870. Quoted in G. E. Bentley Jr., *Blake Records* (Oxford: Clarendon Press, 1969), 221.

177 *"harps which I hear"*: Letter of January 27, 1804. Quoted in Bentley, *Stranger*, 243–44.

177 *dictated to him by a faerie*: Bentley, *Stranger*, 151.

177 *"the madmen outside"*: Quoted in Gilchrist, *Life*, 1:326.

178 *"injured his reputation"*: Henry Crabb Robinson, "William Blake, Künstler, Dichter, und Religiöser Schwärmer," *Vaterländisches Museum* (1811), quoted in Bentley, *Stranger*, 339.

178 *"till I am frightened at it"*: Quoted in A. H. Palmer, *The Life and Letters of Samuel Palmer: Painter and Etcher* (London: Seely and Co., 1892), 24.

178 *"source and nurse of creativity"*: Doris Lessing, "Sketches from Bohemia," *The Guardian*, June 14, 2003.

178 *"live to be hanged"*: The original anecdote is told in Gilchrist, *Life*, 13, but certain details, such as how much Ryland charged to take on apprentices, are added in the retelling in Bentley, *Stranger*, 30–31.

179 *was indeed hanged*: Bentley, *Stranger*, 31.

179 *"worm-lines"*: Ibid., 35. This section in Bentley also has a striking sensory description of the sights, sounds, and smells of an engraver's shop.

179 *"draw a feather over it"*: Quoted in ibid., 35.

180 *"tools of his trade"*: Ibid., 47.

180 *"perverse invention"*: Ibid., 54.

180 *"Joseph, the sacred carpenter"*: Gilchrist, *Life*, 1:69–70. Emphasis original.

180 *cut down on the costs*: Bentley, *Stranger*, 93.

181 *clapping for joy*: Gilchrist, *Life*, 1:59. Gilchrist calls this "a truly Blake-like detail."

181 *"with his spirit I converse"*: Quoted in Bentley, *Stranger*, 21.

181 *rebirth in the art of illuminated manuscripts*: Joseph Viscomi, "Illuminated Printing," in *The Cambridge Companion to William Blake*, ed. Morris Eaves (Cambridge, UK: Cambridge University Press, 2006), 47.

181 *went for almost a million dollars*: The famous Sotheby's sale, *The Library of Abel E. Berland Part I: Important English Literature, Science and Philosophy*, October 8, 2001.

182 *"combines the painter and the poet"*: Quoted in Gilchrist, *Life*, 2:263.

182 *366 leaves*: D. F. McKenzie, *Bibliography and the Sociology of Texts* (Cambridge, UK: Cambridge University Press, 1999), 59. McKenzie cites his reliance on Dr. John Kidd's work for this section.

182 *"blood down Palace walls"*: William Blake, "London," in the Yale Center for British Art's copy L of *Songs of Innocence and of Experience* (1795), object 51.

184 *illustrative act of charity*: Essick, *Huntington*, 60.

184 *we miss major components*: This was actually done as a sort of pedagogical experiment by Jennifer Phegley, as recounted in "They Are Not Just Big, Dusty Novels': Teaching *Hard Times* within the Context of *Household Words*," in *Teaching Bibliography, Textual Criticism, and Book History*, ed. Ann R. Hawkins (London: Pickering and Chatto, 2006), 122–31. Phegley discovered her students were much more sympathetic to the plight of Dickens's characters, and more likely to be convinced by the arguments of his narrative when reading *Hard Times* as part of *Household Words*.

184 *"merest glimpse of that complex world"*: McGann, *Textual*, 33.

184 *"composite art"*: This term was first used by Jean Hagstrum in *William*

Blake: Poet and Painter (1964), but was explored in detail by W. J. T. Mitchell in *Blake's Composite Art* (1978). The concept itself took wing thanks to Northrop Frye, *Fearful Symmetry* (1947).

184 *"hovered over his head"*: Bentley, *Records*, 470–71.

186 *"almost hopeless torment"*: Bentley, *Stranger*, 150.

187 *the Serpent*: Ibid., 150–51.

187 *"sew me together yourself"*: Shelley Jackson, *Patchwork Girl: Or, a Modern Monster* (Watertown, MA: Eastgate Systems, 1996).

187 *"conceive of gender and identity"*: George Landow, "Stitching Together Narrative, Sexuality, Self: Shelley Jackson's Patchwork Girl," review of *Patchwork Girl*, by Shelley Jackson, *electronic book review* (blog), September 1, 1996, http://www.electronicbookreview.com/thread/writingpostfeminism/piecemeal.

187 *"God of this world"*: Annotation by Blake in his copy of Thornton, *Lord's Prayer*, quoted in Bentley, *Stranger*, 9–10.

188 *"gentleman . . . Lord Chancellor"*: Annotation by Blake in his copy of Bacon's *Essays*, quoted in Bentley, *Stranger*, 10.

188 *"I longed to see Satan"*: Allan Cunningham, "William Blake," *The Lives of the Most Eminent British Painters, Sculptors, and Architects, Second Edition* (London: John Murray, 1830), 174.

189 *"deadly terrors"*: William Blake, "The Tyger," in the Yale Center for British Art's copy L of *Songs of Innocence and of Experience* (1795), object 36.

191 *"'scaly skin of gold and green'"*: Cunningham, "Blake," 173.

192 *precise relief etching technique*: See Viscomi, *Idea of the Book*, for the full treatment.

192 *"mere drudgery"*: Letter of Blake, January 1802, quoted in G. E. Bentley Jr., *William Blake in the Desolate Market* (Montreal: McGill-Queen's University Press, 2014), 10.

192 *"tangible link to all the succeeding moments"*: G. Thomas Tanselle, *Bibliographical Analysis: A Historical Introduction* (Cambridge, UK: Cambridge University Press, 2009), 2.

192 *"infinite which was hid"*: William Blake, *The Marriage of Heaven and Hell*, in the Library of Congress's copy D (1785), object 14.

192 *"so evidently insane"*: This was actually written by the famous romantic poet Robert Southey, who was poet laureate for thirty years. Quoted in Bentley, *Stranger*, 341.

193 *"most thoroughly sane man"*: Palmer, *Life*, 23.

194 *"I was Socrates"*: Robinson, *Reminiscences*, 2:301–2.

194 *"most drastic act of reinterpretation"*: Aileen Ward, "William Blake and His Circle," in *Cambridge Companion to William Blake*, ed. Morris Eaves (Cambridge, UK: Cambridge University Press, 2003), 33.

194 *"triumphs of the engraver's art"*: Bentley, *Stranger*, 424.

194 *"chronic copper intoxication"*: Joseph Viscomi, "Blake's Death," *Blake*, vol. 30 (1996): 37.

194 *"use our imaginations"*: Bentley, *Stranger*, 438.

194 *"Listen to the dead with your eyes"*: This quote as applied to the study of books was borrowed from Roger Chartier, *The Author's Hand and the Printer's Mind*, trans. Lydia G. Cochrane (Cambridge, UK: Polity Press, 2014).

Chapter 8: The Memoir That Killed Her Memory

195 *"deviation from rectitude"*: Mary Wollstonecraft, *Posthumous Works of the Author of A Vindication of the Rights of Woman*, 4 vols. (London: J. Johnson, 1798), 3:12.

195 that pen of yours: Cf. Sandra M. Gilbert and Susan Gubar, *The Madwoman in the Attic: The Woman Writer and the Nineteenth-Century Literary Imagination*, 2nd ed. (New Haven, CT: Yale Nota Bene, 2000), which famously begins, "Is a pen a metaphorical penis?" and then systematically demonstrates, with damning quotes from the likes of Ruskin, that, yes, it is indeed a metaphorical penis.

196 *made-up career*: The career "professional author" is defined in the sense of earning pay for writing. See Andrew Bennett, *The Author* (London: Routledge, 2005), 53; and Betsy A. Schellenberg, "The Professional Female Writer," in *The Cambridge Companion to Women's Writing in Britain: 1680–1789*, ed. Catherine Ingrassia (Cambridge, UK: Cambridge University Press, 2015), 48–49.

196 *number of titles*: Michelle Levy, "Women and Print Culture, 1750–1830," in *The History of British Women's Writing, 1750–1830*, vol. 5, ed. Jacqueline M. Labbe (London: Palgrave Macmillan, 2010), 31.

196 *unique genius*: Bennett, *Author*, 52.

196 *landmark court cases*: See Mark Rose, "The Author in Court: *Pope v. Curll* (1741)," *Cultural Critique* 21 (1992): 197–217.

196 *"Mother of Feminism" . . ."Prostitute"*: Susan J. Wolfson, "Mary Wollstonecraft and the poets," in *The Cambridge Companion to Mary Wollstonecraft*, ed. Claudia L. Johnson (Cambridge, UK: Cambridge University Press, 2002), 181.

197 *"thwarting self-murders"*: Janet Todd, *Mary Wollstonecraft: A Revolutionary Life* (rev. ed; London: Bloomsbury Reader, 2014), 356.

197 *"weak as the weakest girl"*: Sylvanus Urban, *Gentlemen's Magazine, and Historical Chronicle* 68, no. 1 (1798): 368. Also quoted in Todd, *Wollstonecraft*, 356–57.

197 *more a slave than a spouse*: Todd, *Wollstonecraft*, 5.

198 *"pedagogue with her own theory of education"*: Ibid., 58.

198 *rock star exceptions such as . . . Christine de Pizan*: Shailor, *Medieval*, 92.

199 *useless liars*: See *Republic* III. In the classical period, the two most common roles of an author were imitation and divine inspiration. Plato makes a distinction between "imitation" and personal narration, and seems okay with poetry in defense of "justice," but he also calls for extensive censorship: "we must put a stop to such stories, lest they produce in the youth a strong inclination to do bad things." *Plato: Complete Works*, ed. John M. Cooper (Indianapolis, IN: Hackett, 1997), 1029. See also Seán Burke, "Changing Conceptions of Authorship," in *Authorship: From Plato to Postmodernism, a Reader*, ed. Seán Burke (Edinburgh: Edinburgh University Press, 1995), 5.

199 *dipped in excrement*: Technically, this is the punishment for flatterers, a subspecies of fraud. We are open to a discussion of whether the punishment for thieves (spontaneous combustion via snakebite) or falsifiers (leprosy and other diseases) might be more suitable. *Inferno*, cantos 17–30.

199 *Equal-magnitude gifts*: *Scientia donum dei est, unde vendi non potest* ("Knowledge is a gift of God, so that it cannot be sold"). Quoted in Burke, *Knowledge*, 149.

199 *paid him for his work*: S. H. Steinberg and John Trevitt, *Five Hundred Years of Printing* (London: British Library, 1996), 108.

199 *"just one of the numerous craftsmen"*: Martha Woodmansee, "On the Author Effect: Recovering Collectivity," in *The Construction of Authorship: Textual Appropriation in Law and Literature*, ed. Martha Woodmansee and Peter Jaszi (Durham, NC: Duke University Press, 1994), 15.

199 *Alexander Pope and Samuel Johnson*: Rose, "Court," 212; and Loewenstein, *Author's Due*, 239.

200 *"authorship" overlaps "authority"*: Bennett, *Author*, 6.

200 *accused of plagiarism*: Catherine Ingrassia, Introduction to *The Cambridge Companion to Mary Wollstonecraft*, ed. Claudia L. Johnson (Cambridge, UK: Cambridge University Press, 2002), 5.

200 *"They'll say it's stoln"*: Quoted in Anne Bradstreet, *The Works of Anne Bradstreet in Prose and Verse*, ed. John Harvard Ellis (Charlestown, MA: Abram E. Cutter, 1867), 101. Originally published in the prologue of her 1650 collection *The Tenth Muse.*

200 *the actual writer of* To Kill a Mockingbird: Glynnis MacNicol, "Harper Lee: The 'Great Lie' She Didn't Write Mockingbird Rears Its Head Again," *The Guardian*, July 20, 2015, http://www.theguardian.com/books/2015/jul/20/harper-lee-to-kill-a-mockingbird-authorship-women-writers.

200 *"denied so as to be attained"*: Seán Burke, "Introduction to Feminism and the Authorial Subject," in *Authorship: From Plato to Postmodernism, a Reader*, ed. Seán Burke (Edinburgh: Edinburgh University Press, 1995), 146.

200 *Mary Brunton*: Brunton's first novel, *Self-Control* (1811), was popular enough to merit three editions in a single year. It wasn't universally liked, however; a contemporary female reader noted, "I read Self-Control and like it extremely, all except some vulgarity meant to be jocular which tired me to death, but I think the principal character charming & well supported & the book really gives good lessons." Is this one of our Amazon reviews? Quoted in Lorna J. Clark, ed., *The Letters of Sarah Harriet Burney* (Athens: University of Georgia Press, 1997), 133.

200 *the Great Forgetting*: Coined by Clifford Siskin in *The Work of Writing: Literature and Social Change in Britain, 1700–1830* (Baltimore, MD: Johns Hopkins University Press, 1998).

201 *"struggle for authorship"*: Burke, "Feminism," 145.

201 *nineteenth-century literary life*: Joanne Shattock, "The Construction of the Woman Writer," in *Women and Literature in Britain, 1800–1900*, ed. Joanne Shattock (Cambridge, UK: Cambridge University Press, 2001), 9.

201 *"great genius"*: Elizabeth Gaskell, *Life of Charlotte Brontë*, 2 vols. (London: Smith, Elder and Co., 1857), 1:66

202 *"same human rights"*: Anne K. Mellor, "Mary Wollstonecraft's *Vindication of the Rights of Women* and the Women Writers of Her Day," in *The Cambridge Companion to Mary Wollstonecraft*, ed. Claudia L. Johnson (Cambridge, UK: Cambridge University Press, 2002), 141.

202 *"same method as boys"*: Ibid., 142.

202 *"always relative to the men"*: Rousseau, *Émile* (1762), quoted in Mary Wollstonecraft, *A Vindication of the Rights of Women*, 2nd ed. (London: J. Johnson, 1792), 175.

203 *"best and most important"*: Jean-Jacques Rousseau, *Confessions*, trans. J. M. Cohen (New York: Penguin, 1953), 529–30.

203 *"diametrically opposite"*: Wollstonecraft, *Vindication*, 83.

203 *complications from childbirth*: Godwin's memoir contains a graphic play-by-play of Wollstonecraft's agonizing death from septicaemia (in which the placenta rots inside the mother rather than being expelled), on pages 181–98. A particularly heartbreaking passage: "He immediately proceeded to the extraction of the placenta, which he brought away in pieces, till he was satisfied

that the whole was removed. In that point however it afterwards appeared that he was mistaken." Godwin, *Memoirs*, 181.

204 read her works and not also fall in love: Of Wollstonecraft's *Letters from Sweden*, Godwin once remarked, "If ever there was a book calculated to make a man in love with its author, this appears to me to be the book." Quoted in Todd, *Wollstonecraft*, 369.

204 "marks of contempt": Godwin, *Memoirs*, 9.

204 "active and hardy sports": Ibid., 12–13.

205 discredited . . . as a result: Take the previously mentioned Enlightenment philosopher Rousseau, who not only couldn't keep himself from making five aftermaths out of wedlock, but then, in order to "preserve the honor of the mother" (mocking air quotes, there, not actual quotes) convinced her to give them all up for adoption. And this was the author who became an international celebrity for his treatises on education and child rearing.

205 "Unsex'd female": Richard Polwhele, *The Unsex'd Females: A Poem* (London: Cadell and Davies, 1798).

205 "challenging men to duels": Todd, *Wollstonecraft*, 185.

205 "apt to wreck": Anna Katherine Elwood, *Memoirs of the Literary Ladies of England*, 2 vols. (London: Henry Colburn, 1843), 2:152.

205 earliest sympathetic discussions: Shattock, "Construction," 15.

207 "alum and soap": Todd, *Wollstonecraft*, 267.

207 nearly devoid of sex: This can be awkward for the likes of second-wave feminists, who sometimes "brand her a sexual puritan." See Barbara Taylor, *Mary Wollstonecraft and the Feminist Imagination* (Cambridge, UK: Cambridge University Press, 2003), 117.

207 "personal intercourse of appetite": Wollstonecraft, *Vindication*, 222.

207 "sexual relationship outside marriage": Todd, *Wollstonecraft*, 230.

207 hunger or thirst: Ibid., 235.

208 "sexual desire as natural and right": Ibid., 236.

208 "continued to detain him": Godwin, *Memoirs*, 116.

208 presumably laudanum: Todd, *Wollstonecraft*, 287.

209 "morality above literary ambition": Ingrassia, Introduction, 2–3.

209 "propriety, modesty, and decorum": Ibid.

209 "propagating w----s": *The Anti-Jacobin Review* 9 (London: Anti-Jacobin Office, 1801), 518.

210 Prostitution *was cross-referenced*: Index of *The Anti-Jacobin Review* 1

(London: Anti-Jacobin Office, 1798), though we also notice these guys were still making the joke as late as 1890, in the *Index of a Compilation of the Poetry of the Anti-Jacobin*, ed. Charles Edmonds (London: Sampson Low, Marston, Searle, and Rivington, 1890), 339.

210 *emotional "scribblers"*: Cf. Nathaniel Hawthorne's comment, "America is now wholly given over to a d—d mob of scribbling women, and I should have no chance of success while the public taste is occupied with their trash," in a letter to William D. Ticknor, January 19, 1855, in *Nathaniel Hawthorne, The Letters, 1853–1856*, ed. Thomas Woodson et al., *The Centenary Edition of the Works of Nathaniel Hawthorne*, vol. 16 (Columbus: Ohio State University. Press, 1987), 304. See also Michael Winship, "Hawthorne and the 'Scribbling Women': Publishing *The Scarlet Letter* in the Nineteenth-Century United States," *Studies in American Fiction* 29, no. 1 (2001): 3–11.

210 *"identify openly with Wollstonecraft"*: Mellor, "Women Writers," 145.

210 *"tarred and feathered"*: Ibid.

211 *overlooked details of Wollstonecraft's life*: Shattock, "Construction," 16–17.

211 *New editions of* Vindication: Ibid., 18.

211 *resurrected and reinvented*: Cora Kaplan, "Mary Wollstonecraft's Reception and Legacies," in *The Cambridge Companion to Mary Wollstonecraft*, ed. Claudia L. Johnson (Cambridge, UK: Cambridge University Press, 2002), 247–49. This article provides an excellent summary of the vicissitudes of critical opinion on Wollstonecraft.

211 *more a reflection of our society*: In the 1968 essay "What Is an Author?," Michel Foucault famously declared the author nothing but an artificial construct, "a certain functional principle by which, in our culture, one limits, excludes, and chooses." Each culture defines these boundaries in different ways, which leads to a different interpretation of what an author is. Michel Foucault, "What Is an Author?'" in Seán Burke, *Authorship: From Plato to Postmodernism, a Reader*, ed. Seán Burke (Edinburgh: Edinburgh University Press, 1995), 125–30.

211 *which is to say, forever*: A postmodernist statement that reflects Roland Barthes's assertion that readers are authorities over any concrete idea of the author. According to Barthes, "To give a text an Author is to impose a limit on that text." Roland Barthes, "The Death of the Author," in Burke, *Authorship*, 129. Most scholars today believe that the idea of the author as a unique talent does not independently exist. In the very least, authorship is considered much more collaborative than has historically been the case; see *Textual Appropriation in Law and Literature*, ed. Martha Woodmansee and Peter Jaszi (Durham, NC: Duke University Press, 1994), especially the introduction by Woodmansee and Jaszi.

211 *"act of defiance"*: Dale Spender, *Mothers of the Novel: 100 Good Women*

Writers before Jane Austen (London: Pandora, 1986), 3. See also Cheryl Turner, *Living by the Pen: Women Writers in the Eighteenth Century* (London: Routledge, 2002), 10–11.

212 thought half as good: P. T. Rooke and R. L. Schnell, *No Bleeding Heart: Charlotte Whitton, A Feminist on the Right* (Vancouver: University of British Columbia Press, 1987), 2.

Chapter 9: American Bookaneers

213 "special production": Lord Shaftesbury, diary entry for December 20, 1871, quoted in Edwin Hodder, *The Life and Work of the Seventh Earl of Shaftesbury, K.G.* (London: Cassell and Company, 1887), 658.

214 "emperors and kings": Sidney Phil Moss, *Charles Dickens' Quarrel with America* (Albany, NY: Whitston Publishing Company, 1984), 2.

214 "never been so feted": Ibid.

214 "heart is democratic": Quoted in Jerome Meckier, *Innocent Abroad: Charles Dickens' American Engagements* (Lexington: University Press of Kentucky, 1990), 16.

214 "Here we are!": Recounted in Michael Slater, *Charles Dickens: A Life Defined by Writing* (New Haven, CT: Yale University Press, 2009), 178.

214 "I would have the whole United States": Quoted in ibid., 180.

214 "penny-a-liner loafer": Quoted in Moss, *Quarrel*, 40.

215 "silliness and inanity": Quoted in ibid.

215 "hangs up his fiddle": Quoted in ibid., 133.

215 "more popular than Jesus": Lennon's remarks went without comment when published in the *London Evening Standard*. It was when they were published five months later in a U.S. magazine, *Datebook*, that the uproar began in the American South. See Jonathan Gould, *Can't Buy Me Love: The Beatles, Britain, and America* (New York: Three Rivers Press, 2007), 309, 341–43.

215 restricted to American citizens: James J. Barnes, *Authors, Publishers, and Politicians: The Quest for an Anglo-American Copyright Agreement, 1815–1854* (London: Routledge, 1974), 50.

216 "older generation deplored": Robert L. Patten, *Charles Dickens and His Publishers* (Santa Cruz: University of California Press, 1978), 10.

216 "an author in every fifteenth or twentieth person": Florian Schweizer, "Authorship and the Professional Writer," in *Charles Dickens in Context*, ed. Sally Ledger and Holly Furneaux (Cambridge, UK: Cambridge University Press, 2011), 119.

216 Lord Byron and Sir Walter Scott: Ibid.; and Patten, *Publishers*, 21.

216 *source of depravity*: Peter Baldwin, *The Copyright Wars: Three Centuries of Trans-Atlantic Battle* (Princeton, NJ: Princeton University Press, 2014), 7.

216 *"I'm lovin' it"*: The company really does have this trademarked. On the McDonald's official website is a section listing its myriad trademarks, including "Chicken McNuggets . . . Egg McMuffin, Extra Value Meal . . . Fish McBites . . . Golden Arches Logo, Good Time, Great Taste . . . Triple Thick, twoallbeefpattiesspecialsaucelettucecheesepicklesonionsonasesameseedbun, We Love to See You Smile," and many, many more. See http://www.mcdonalds.com/us/en/terms_conditions.html.

217 *"libel, blasphemy, or sedition"*: Mark Rose, "Author in Court," 201.

217 *legal entity*: Ibid., 199–202.

217 *"double the sum"*: Meckier, *Innocent*, 55.

218 *"baracoon of 'bookaneers'"*: *The Metropolitan*, March 1853, 86. This was actually an American monthly magazine arguing that U.S. piracy hurt homegrown authors. The Hood reference is to Thomas Hood, the respected English poet. He first used the term in an 1837 letter printed in the *Athenaeum* entitled "Copyright and Copywrong."

218 *piracy = educating*: Barnes, *Quest*, 15; and Baldwin, *Copyright Wars*, 114.

219 *Dickens cap-shouted*: Alexander Welsh, *From Copyright to Copperfield: The Identity of Dickens* (Cambridge, MA: Harvard University Press, 1987), 9.

220 *roaming passels of hogs*: Slater, *Dickens*, 190.

220 *"most photographically famous"*: Joss Marsh, "The Rise of Celebrity Culture," in Ledger and Furneaux, *Charles Dickens in Context*, 106.

220 *"furiously walking, in flamboyant clothes"*: Ibid., 104.

220 *"admiration of a Kentucky huntsman"*: Report of a journalist at the *Worcester Aegis* during Dickens's visit, quoted in Slater, *Dickens*, 181.

220 *"graphic fancy"*: Ibid.

220 *mistaken for a riverboat gambler*: Meckier, *Innocent*, 23.

220 *"pioneer first-class hotel"*: Slater, *Dickens*, 178.

220 *"straight down upon [Niagara] Falls"*: Ibid., 187.

221 *£5,019 in debt*: Meckier, *Innocent*, 58. See also Patten, *Publishers*, 125.

221 *receipt from his publishers*: Slater, *Dickens*, 175.

221 *"minor figures in Victorian letters to titans"*: Patten, *Publishers*, 46.

221 *wasn't financially secure*: Ibid., 10.

221 *three hundred fifty pounds from American publishers*: Moss, *Quarrel*, 112.

221 *modest two thousand pounds*: Patten, *Publishers*, 70.

222 *"lay particular stress"*: William Glyde Wilkins, ed. and comp., *Charles Dickens in America* (London: Chapman and Hall, 1911), 33.

222 *"dwellers in log-houses"*: Ibid., 32.

222 *"substantial profit and return"*: Ibid., 33.

222 *"tumultuous" applause*: Edgar Johnson, *Charles Dickens: His Tragedy and Triumph* (London: Simon and Schuster, 1952), 1:375. Also recounted in Meckier, *Innocent*, 41.

223 *administrative oversight*: Meckier, *Innocent*, 52.

223 *"bad taste" . . . "smells of the shop"*: Moss, *Quarrel*, 10.

223 *"pay homage" . . ."refrains"*: Quoted in Paul B. Davis, "Dickens and the American Press, 1842," *Dickens Studies* 4 (1968), 68–69.

223 *"whisper in your ear"*: Quoted in Welsh, *Copyright*, 32.

224 *"one grateful dollar"*: Quoted in Meckier, *Innocent*, 44.

224 *"pressure on his brain"*: Quoted in Slater, *Dickens*, 182.

224 *"indelicacy and gross impropriety"*: Quoted in Moss, *Quarrel*, 3.

224 *"worst taste possible"*: Ibid., 10.

225 *"1839 panic set in"*: Barnes, *Quest*, 2.

225 *"fell lower and lower"*: Ibid., 1.

225 *three-quarter price drop*: Ibid., 4.

225 *"than most literary periodicals"*: Ibid., 8.

225 *weeklies . . . "custom of reprinting"*: See Catherine Seville, *The Internationalisation of Copyright Law: Books, Buccaneers and the Black Flag in the Nineteenth Century* (Cambridge, UK: Cambridge University Press, 2006), 168.

226 *"owes us a debt of gratitude"*: Quoted in Barnes, *Quest*, 15.

226 *worst copyright offenders*: Baldwin, *Copyright Wars*, 113.

226 *"enlightened and democratic polity"*: Ibid., 114.

226 *railway timetables*: Ibid., 117.

226 *three to four times as many books*: Ibid., 118.

226 *"purportedly made the writer famous"*: Meckier, *Innocent*, 42.

227 *"water closet paper"*: From 1880 letter from Twain to William Dean Howells, quoted in Siva Vaidhyanathan, *Copyrights and Copywrongs: The Rise of Intellectual Property and How It Threatens Creativity* (New York: New York University Press, 2001), 60.

227 *cost four times as much*: Baldwin, *Copyright Wars*, 119.

227 *"honestly come by"*: Quoted in ibid., 120, but originally from Report 1188, May 21, 1886, of the *Reports of the Committees of the Senate.*

227 *"country is drugged"*: Quoted in ibid., 120, but originally from Report 622, March 19, 1888, of the *Reports of Committees of the Senate.*

227 *"dark, slimy, universal pond"*: "Address to the People of the United States in Behalf of the American Copyright Club, Adopted at New-York, October 18th, 1843" (New York: American Copyright Club, 1843), 11.

227 *"colonial bondage"*: From an 1842 letter from Samuel Morse to Cornelius Mathews, one of the leading members of the American Copyright Club. Quoted in Barnes, *Quest*, 83.

227 *"clamor of two hundred authors"*: Quoted in Baldwin, *Copyright Wars*, 119, but originally from Report 1188, May 21, 1886, of the *Reports of the Committees of the Senate.*

228 *"Republic of my imagination"*: Quoted in Meckier, *Innocent*, 19, who describes it as "the saddest traveler's letter ever penned."

228 *"left off washing himself"*: Charles Dickens, *The Life and Adventures of Martin Chuzzlewit* (London: Chapman and Hall, 1844), 398.

229 *solemn and an awful thing*: Ibid., 204–5.

229 *"offend and disgust another"*: Ibid., 403.

229 *witty little observer*: Meckier, *Innocent*, 6.

229 *peer inside the cabin*: Quoted in ibid.

229 *hair clippings for sale*: Ibid.

229 *"looking down my throat"*: Quoted in ibid., 7.

229 *"'Teaberry Tooth Wash'"*: Slater, *Dickens*, 195.

230 *"Boz Pork & Beans"*: Ibid.

230 *"common, decent, natural, human politeness"*: Dickens, *Chuzzlewit*, 403.

231 *"great bond of equality"*: Slater, *Dickens*, 185.

231 *"intolerably conceited"*: Ibid., 186.

231 *"sublimity of nature"*: Ibid.

231 *bordered "on the ridiculous"*: Moss, *Quarrel*, 82.

231 *"sacred wrath of the newspapers"*: John Forster, *The Life of Charles Dickens*, 3 vols. (London: Chapman and Hall, 1872–1874), 1:326.

232 *"have to denounce as slanderous"*: Meckier, *Innocent*, 56.

233 *imported books face duties*: Ibid., 72.

233 *"to our own wants"*: Moss, *Quarrel*, 100.

233 *they'd already been doing*: See ibid.

234 *"simultaneously published"*: N. N. Feltes, "International Copyright: Structuring 'The Condition of Modernity' in British Publishing," in *The Construction of Authorship: Textual Appropriation in Law and Literature*, ed. Martha Woodmansee and Peter Jaszi (Durham, NC: Duke University Press, 1994), 272.

234 *"eminently comprehensible to those nineteenth-century reformers"*: Baldwin, *Copyright Wars*, 13.

Chapter 10: When Doves Cry

235 *"my last cards"*: T. J. Cobden-Sanderson, *The Journals of T. J. Cobden-San derson 1879–1922*, 2 vols. (New York: Macmillan, 1926), 2:213.

236 *"work and disappear"*: This is actually a famous line by Robert Bringhurst about typographers themselves, but the principle still applies in line with his intent. Robert Bringhurst, *The Elements of Typographic Style*, 4th ed. (Seattle: Hartley and Marks, 2012), 21.

236 *obvious rip-off*: See Simon Garfield, *Just My Type: A Book About Fonts* (London: Profile Books, 2011), 221. "Arial is the Helvetica lookalike . . . still regarded—and rightly so—as a cheat."

236 *most recognizable, truly American font*: Ibid., 201–2.

236 *poor Comic Sans*: Ibid., 20. One of the comic books that inspired Vincent Connare, the creator of the typeface, was *Batman: The Dark Knight Returns*, by Frank Miller and Klaus Janson.

236 *steel-toe boots send a very different message*: Erik Spiekermann, *Stop Stealing Sheep and Find Out How Type Works*, 3rd ed. (San Francisco, CA: Peachpit, 2014), 42–45.

236 *"unremarkability and inoffensability" . . ."frightening the horses"*: Garfield, *Type*, 211.

237 *more likely to agree*: Errol Morris, "Hear, All Ye People; Hearken, O Earth (Part 2)," *New York Times*, August 9, 2012, http://opinionator.blogs.nytimes.com/2012/08/09/hear-all-ye-people-hearken-o-earth-part-2/?_r=0

238 *"at the same time something beautiful"*: Quoted in Roderick Cave, *The Private Press* (New Providence, NJ: R. R. Bowker Company, 1983), 120–21.

238 *counterfeit Doves bindings*: See Marianne Tidcombe, *The Doves Bindery* (London: British Library, 1991), 458–65.

238 *moving belt and felt rollers*: Arthur Chick, *Towards Today's Book: Progress in 19th Century Britain* (London: Farrand Press, 1997), 21–22.

239 *machine-made paper*: James Mosley, "The Technologies of Print," in *The*

Book: A Global History, ed. Michael F. Suarez, S.J., and H. R. Woudhuysen (Oxford: Oxford University Press, 2013), 145.

239 *newly mechanized presses*: Chick, *Today's Book*, 36–38.

239 *more than tripled its productivity*: Mosley states that hand press technology at the *Times* "could not exceed 300 impressions in an hour," while the speed of the early Koenig presses was 1,000 impressions an hour. Mosley, "Technologies," 160.

239 *mechanized printing assembly line*: Ibid., 145.

239 *newspapers were no longer constrained*: Lyle L. Miller, *Maintaining Reading Efficiency* (New York: Henry Holt and Company, 1959), 34.

239 *specialized typecasting machines*: Mosley, "Technologies," 148.

239 *"vending machine"*: Bringhusrt, *Elements*, 137.

239 *fast, cheap, or good*: This is a common riff on the Iron Triangle, a tool in project management first introduced by Martin Barnes in 1969. These days, however, it tends to be updated to the Project Diamond, which acknowledges that a multitude of factors plays into these three aspects.

240 *for the beauty of their craft*: Cave points to Bernard von Mallinckrodt's 1639 book, *De Ortu ac Progressu Artis Typographiae*, as the first to reference private presses in literature. See Cave, *Private*, 3.

240 *"distinct undercurrent"*: Ibid., 4.

240 *transformed into a full-fledged rebellion*: Stanley Morison, *A Tally of Types*, with new introduction by Michael Parker (Boston: David R. Godine, 1999), 17.

240 *William Morris, the Pre-Raphaelite poet and textile designer*: Morris is, obviously, the Obi-Wan of the Private Press Movement.

240 *first book of the Private Press Movement*: Colin Franklin, *The Private Presses*, 2nd ed., with bibliography and indexes by John Turner (London: Scolar Press, 1991), 9.

241 *"state of the nation"*: Ibid., 43.

241 *"advice and approval"*: Quoted in Cave, *Private*, 104. In correspondence with the authors, Robert Gree suggested the "silent partner" description.

241 *food be damned*: See ibid., xviii.

242 *18 percent of its total start-up cost*: Tidcombe, *Doves*, 223. Cobden-Sanderson was apparently "delighted" to get that much, having expected he would have gotten closer to forty pounds.

242 *"The Book Beautiful"*: See T. J. Cobden-Sanderson, *The Ideal Book or Book Beautiful* (Hammersmith, UK: Doves Press, 1900).

242 *"sacred vessels of Western culture"*: Megan L. Benton, *Beauty and the Book:*

Fine Editions and Cultural Distinction in America (New Haven, CT: Yale University Press, 2000), 4.

242 *"a dream, a symbol of the infinitely beautiful"*: Cobden-Sanderson, *Ideal*, 9.

242 *"I know no more perfect book"*: A. W. Pollard, "The Doves Press," in *Cobden-Sanderson and the Doves Press*, ed. John Henry Nash (San Francisco: John Henry Nash, 1929), 13.

243 *language of the Press*: T. J. Cobden-Sanderson, *Cosmic Vision* (Thavies Inn, [London]: Richard Cobden-Sanderson, 1922), 124. It first appeared as "Notes on the Doves Press" in the 1908 Doves *Catalogue Raisonné.*

244 *"only another printer can recognize"*: Cave, *Private*, 122.

244 *"profits after expenses"*: Marianne Tidcombe, *The Doves Press* (London: The British Library, 2002), 28.

245 *personal touch to the numerals*: Ibid., 23.

245 *way for scribes to write faster*: Shailor, *Medieval Book*, 28.

246 *"signify the classicist's contempt"*: E. P. Goldschmidt, *The Printed Book of the Renaissance: Three Lectures on Type, Illustration and Ornament* (Cambridge, UK: Cambridge University Press, 2010), 2.

246 *Renaissance-era insult*: In a lovely bit of irony, the Renaissance authors who dismissed nonclassical Gothic scripts weren't aware that the scripts they themselves most revered, and which they believed dated to antiquity, were in fact Carolingian, that is, smack in the middle of the Middle Ages.

246 *"be German, even in your script"*: Quoted in Garfield, *Type*, 191.

246 *"Jewish owners"*: Ibid.

246 *shortage of type in Gothic fonts*: Quoted in ibid., 192.

247 *"full of wine"*: Charles Ricketts, *A Defence of the Revival of Printing* (London: Vale Press, 1899), 19.

247 *"alienation effect"*: McGann, *Textual*, 85.

247 *Sotheby's auction*: Tidcombe, *Doves Press*, 13.

247 *"absolutely perfect"*: Quoted in ibid., 22.

248 *"over-inked"*: Ibid., 14–15.

249 *"death's door"*: Unsent letter from Cobden-Sanderson to Emery Walker, August 2, 1902. Part of Cobden-Sanderson's *Pro Iracundia sua Apologia*, his extended defense of his destruction of the Doves Type, recorded in full in Tidcombe, *Doves Press*, 103–4.

250 *"said it was 'hateful'"*: Ibid.

251 *"watch plants and insect forms"*: Franklin, *Private*, 115.

251 *"cut it, and cut it, and slashed it"*: Quoted in ibid., 117.

251 *fine press blasphemy*: According to John Mason, a compositor at the Doves Press, Cobden-Sanderson "regarded [the type] as a consecrated instrument, and shrank away from what he regarded as desecration," especially the possible use of the type for commercial purposes. Quoted in Leslie T. Owens, *J. H. Mason 1875–1951, Scholar-Printer* (London: Frederick Muller Limited, 1976), 37.

252 *"a Visionary and a Fanatic"*: Letter from Cobden-Sanderson to Sidney Cockerell, June 14, 1909, quoted in Tidcombe, *Doves Press*, 127.

253 *worth less than its original investment*: Ibid., 59.

253 *upon the advice of friends*: Ibid.

253 *"hand and arm of man or woman"*: February 9, 1909, diary entry in Cobden-Sanderson, *Journals*, 2:138.

253 *"Bed of the River Thames"*: June 11, 1911, diary entry in ibid., 2:181–82. *Fount* is the technical term for the physically cast type itself.

253 *cast into the English Channel*: See Cave, *Private*, 324.

253 *"stale by unthinking use"*: Charles Ricketts, *A Bibliography of Books Issued by Hacon and Ricketts* (London: Charles Ricketts, 1904).

254 *destruction of the Doves Type*: See Tidcombe, *Doves Press*, 76.

254 *"My folly is of a light kind"*: October 28, 1916, diary entry in Cobden-Sanderson, *Journals*, 2:301.

254 *"destroyed the whole of it"*: August 12, 1916 (midnight), diary entry in ibid., 2:296.

254 *2,600 pounds of metal type*: Tidcombe, *Doves Press*, 78.

255 *170 individual trips*: Ibid.

255 *"perils and panics"*: November 5, 1916, diary entry in Cobden-Sanderson, *Journals*, 2:303–4.

255 *"lurking in dark corners"*: Ibid.

256 *"the books themselves"*: December 21, 1917, diary entry in ibid., 2:341.

256 *upward of twelve hundred pounds*: We don't know the exact cost. Annie was ordered to pay seven hundred pounds to Walker, but that would not have included lawyers' fees. See Tidcombe, *Doves Press*, 84–85, for a discussion of estimates.

256 *entombed in its waters*: Ibid., 82.

257 *release it as a digital download*: "New Digital 'Facsimile' of Legendary Doves Type: Doves Press Font Revived by Robert Green," *TypeSpec* (blog), http://www.typespec.co.uk/doves-type/.

257 *exhuming 150 individual pieces*: "Recovering the Doves Type," *TypeSpec* (blog), http://www.typespec.co.uk/recovering-the-doves-type/.

257 *digital re-creation of the complete font*: Available at http://www.typespec.co.uk/doves-type/. Buy it. Use it. Love it.

257 *Emery Walker Trust*: Justin Quirk, "X Marks the Spot," *Sunday Times*, January 11, 2015. In correspondence with the authors, Green added, "The other half will go to my children when I die. It's not for sale . . . its sacred."

257 *reexamination of every aspect*: Franklin, *Private*, 14.

258 *twentieth-century book design*: Daniel Berkeley Updike, *Printing Types: Their History, Forms, and Use* (New Castle, DE: Oak Knoll Press, 2001), 208. In correspondence with the authors, Robert Green pointed out that Tschichold's iconic Penguin cover designs, for instance, owe a great deal to the Doves aesthetic philosophy.

258 *few could have predicted*: First there was the influence and dominance (in England, at least) of the Monotype Corporation, and then of course photographic and especially digital printing methods enormously expanded the possibilities of typefaces.

Chapter 11: Blifter!

259 *"MURDER?"*: John Tebbel, *History of Book Publishing in the United States*, vol. 3: *The Golden Age Between the Two Wars (1920–1940)* (New York: Bowker, 1978), 334–35.

259 *"Era of Wonderful Opportunity"*: Ibid., 32.

259 *"sandwich men"*: Ibid., 332.

260 *"needed to be learned"*: Ted Striphas, *The Late Age of Print: Everyday Book Culture from Consumerism to Control* (New York: Columbia University Press, 2009), 27.

260 *owned no books whatsoever*: Benton, *Beauty*, 14.

260 *4 percent of Americans visited a bookstore*: Ibid., 15.

260 *"like furniture"*: Ibid., 14.

261 *"lost touch with supply and demand"*: Striphas, *Late*, 84.

261 *"greatest idea that America has given the world"*: Christine Frederick, *Selling Mrs. Consumer* (New York: The Business Bourse, 1929), 4–5; quoted in Stuart Ewen, *Captains of Consciousness: Advertising and the Social Roots of the Consumer Culture, 25th Anniversary Edition* (New York: Basic Books, 2001), 22.

262 *books you don't really want to read with your dad*: Striphas, *Late*, 29.

262 *Illiteracy . . . and urbanization*: Tebbel, *History*, 4–5.

262 "[The] sign 'Books Wanted'": Ibid., 30.

262 "come into its own": Letter, "To Our Friends in the Trade," printed in *Publishers Weekly*, April 26, 1919, 1168.

263 "very stupid lot": Letter of J. C. Dana, "To the Editor of *The Literary Review*," printed in *Publishers Weekly*, November 27, 1920, 1707–8.

263 "$11.00 for candy": Tebbel, *History*, 65.

263 "soap manufacturer might invest in soap": Quoted in ibid., 315–16; and *Business Digest and Investment Weekly*, May 25, 1920, 681.

263 tailored specifically to the book industry: Tebbel, *History*, 321.

263 devoted to the author himself: See ibid., 48–49 and 327.

263 hyperbolic ascent: Ibid., 327.

263 printed the first chapter of the next: Ibid., 320.

263 "by Billy Sunday": Quoted in ibid., 28.

264 "throw the jacket away": Quoted in ibid., 327. Tebbel remarks that "Phelps' statement was put down to critical petulance and ignored."

264 reissued the book . . . A Prostitute's Sacrifice: Ibid., 207.

264 "100 best novels": Bernays references this with approval: "The release provoked heated discussion and probably stimulated some people to buy the books he mentioned." Edward L. Bernays, *Biography of an Idea* (New York: Simon and Schuster, 1965), 487.

265 Enter the bookshelf: Striphas has an excellent extended section on the bookshelf in this era, upon which this section relies. See "Shelf Life," in *Late*, 26–31.

265 "effect will be pleasing": Quoted in ibid., 27.

265 "devoid of anything within": Quoted in ibid.

265 "perceived changes in the status": Janice A. Radway, *A Feeling for Books: The Book-of-the-Month Club, Literary Taste, and Middle-Class Desire* (Chapel Hill: University of North Carolina Press, 1997), 145.

265 "fetishism of commodities": Ibid., 148. This concept's roots come from the nineteenth-century and the writings of Karl Marx. With thanks to Loren Glass for emphasizing this.

265 "CULTURE, WEALTH, BEHAVIOR, POWER": Joan Shelley Rubin, *The Making of Middlebrow Culture* (Chapel Hill: University of North Carolina Press, 1992), 30.

266 "buyers were hooked into returning": Ibid., 95.

266 well-respected literary judges: Among them was Christopher Morley, the founder of the famous Sherlock Holmes society known as the Baker Street Irregulars and the author of some of the most beloved books on books, such as *Parnassus on Wheels* (1917) and *The Haunted Bookshop* (1919).

266 *cheap prices as its main selling point*: Tebbel, *History*, 295.

267 *book prices had gone up only 50–60 percent*: Ibid., 66.

267 *selling price-slashed books*: Ibid., 312.

268 *"development of its rural areas"*: Larry Tye, *The Father of Spin: Edward L. Bernays and the Birth of Public Relations* (New York: Macmillan, 2002), 118.

268 *"pull the wires"*: Edward L. Bernays, *Propaganda* (New York: Horace Liveright, 1928), 10.

268 *raking in more than $98,000*: Tye, *Spin*, 60. The $1.5 million number comes from the inflation calculator of the Bureau of Labor Statistics, U.S. Department of Labor, http://www.bls.gov/data/inflation_calculator.htm.

269 *"I prefer chocolate"*: Ibid., 27, quoting from a February 2, 1984, interview with the *St. Petersburg Times*.

269 *"crystallizing the obscure tendencies"*: Edward L. Bernays, *Crystallizing Public Opinion* (New York: Boni and Liveright, 1923), 173.

269 *more important than legal advice*: Tye, *Spin*, 60–61, quoting from a 1985 interview with *Forbes*.

269 *"point of view"*: Bernays, *Crystallizing*, 212.

270 *"damn near double our female market" . . ."psychological basis"*: Bernays, *Biography*, 386.

270 *"torches of freedom" . . ."network of media"*: Ibid., 386–87.

271 *"dollar books were not in the public interest"*: Bernays, *Biography*, 485.

271 *"wretch who raised hell with book sales"*: Ibid., 488.

271 *"gain more attention . . . lethal epithet"*: Ibid., 486, 488.

272 *"death of six thousand book retailers"*: Ibid., 485.

272 *"sad figure"*: Ibid.

273 *"Benjamin Franklin . . . the career woman, a relatively new phenomenon"*: Ibid., 487.

273 *first instance of modern American book advertising*: Tebbel, *History*, 27, also referencing a *Post* article from 1919. Whitman did this without Emerson's permission; the famed Transcendentalist was not pleased.

274 *"could not run successfully on this price structure"*: Bernays, *Biography*, 489.

274 *blamed on readers:* Comment made by Alfred Knopf, excerpted in Richard Layman, *Discovering the Maltese Falcon and Sam Spade: The Evolution of Dashiell Hammett's Masterpiece* (San Francisco: Vince Emery, 2005), 166.

275 *"left shivering on the doorstep"*: Bernays, *Biography*, 489.

275 *declared unconstitutional*: Tebbel, *History*, 459.

275 *"bomb"*: Ibid., 439.

276 *"disparaged publishers and editors"*: Striphas, *Late*, 87–88.

276 *revolutionary ISBN system*: Ibid., 83.

276 *"reading can be"*: Earnest Elmo Calkins, 1922, quoted in Tebbel, *History*, 319.

278 *"deliberate, planned campaign"*: All quotes in this anecdote are taken directly from Bernays's own retelling in *Biography*, 652.

Conclusion: Nothing More Deceptive than an Obvious Fact

279 *"books they need are in existence"*: Christopher Morley, *Haunted Bookshop* (New York: Melville House), 18.

279 *painting of Thomas More*: Jonathan Jones, "*Wolf Hall* Is Wrong: Thomas More Was a Funny, Feminist Renaissance Man," *The Guardian*, January 29, 2015.

279 *great portraitists of the Renaissance*: So great a portraitist, in fact, that Henry VIII agreed to marry his fourth wife, Anne of Cleves, on the strength of a portrait executed by Holbein. After Henry met her, he refused to consummate the marriage, claiming he had been misled by the beauty of the portrait.

280 *"more deceptive than an obvious fact"*: Doyle, "Boscombe Valley Mystery," in *Complete Works of Sherlock Holmes*, 1:108.

280 *"integral to More's purpose"*: Duffy, "Defence of Christendom," 200; More, *Dialogue*, 207.

282 *primed for the doubt*: Wilding, *Faussaire*, 56.

283 *"live in an expanding culture"*: Raymond Williams, *Culture and Society, 1780–1950* (New York: Columbia University Press, 1983), viii.

284 *"rewrite history for itself"*: Quoted in Lawrence C. Wroth, *Notes for Bibliophiles in the* New-York Herald Tribune, *1937–1947*, ed. Richard J. Ring (South Freeport, ME: Ascensius Press, 2016), 128.

284 *"felt the case to be a hopeless one"*: Quoted in Meckier, *Innocent*, 68.

INDEX

Entries in *italics* refer to illustrations.

INDEX

INDEX

INDEX

INDEX

ABOUT THE AUTHORS

Rebecca Romney's appearances on the History Channel's *Pawn Stars* have made her the country's most recognizable rare book dealer. In her career, she has sold single volumes for $500,000, books from the libraries of Isaac Newton and Queen Elizabeth I, and more than one Shakespeare Folio.

J. P. Romney is a historical researcher and the author of the young adult novel *The Monster on the Road Is Me.*

ABOUT THE FONT

This book is set in Bodoni, a composite typeface influenced by a number of the historical fonts created by the eighteenth-century Italian type designer Giambattista Bodoni. In his landmark book *Printing Types: Their History, Forms, and Use*, Daniel Berkeley Updike alludes to the typeface's simultaneously playful and elegant style: "a volume set in [Bodoni] suggests a Continental reprint of an English book—an impression by which one is perpetually, though perhaps subconsciously, teased."